Matheus Guimarães Lima

# Ostentatious Funk in Presidente Prudente - SP

Matheus Guimarães Lima

# Ostentatious Funk in Presidente Prudente - SP

ScienciaScripts

**Imprint**
Any brand names and product names mentioned in this book are subject to trademark, brand or patent protection and are trademarks or registered trademarks of their respective holders. The use of brand names, product names, common names, trade names, product descriptions etc. even without a particular marking in this work is in no way to be construed to mean that such names may be regarded as unrestricted in respect of trademark and brand protection legislation and could thus be used by anyone.

Cover image: www.ingimage.com

This book is a translation from the original published under ISBN 978-3-330-74441-7.

Publisher:
Sciencia Scripts
is a trademark of
Dodo Books Indian Ocean Ltd. and OmniScriptum S.R.L publishing group

120 High Road, East Finchley, London, N2 9ED, United Kingdom
Str. Armeneasca 28/1, office 1, Chisinau MD-2012, Republic of Moldova, Europe
Printed at: see last page
**ISBN: 978-620-8-07981-9**

I dedicate it especially to all the funkeiros, DJs and MCs. I dedicate it to the universal language, music.

# ACKNOWLEDGEMENTS

I thank God for the opportunity to have a quality education. I thank my parents Carlos and Eunice for everything. A huge thank you to my advisor Prof Dr Nécio Turra Neto for his support during the development of this project; without his help it wouldn't have been possible. I would also like to thank DJ Robinho's troop: DJ Robinho, MC Dark, MC Nego da Oeste, MC Laurinho LB and MC Etoo SP. The research that resulted in this book would never have happened if it hadn't been for you, thank you all very much for your contribution and long live Funk.

*"Rolex watch, Double X, Ed Hardy, the firm is strong. I get to the shopping centre, hey manager, I want to leave all wearing Oakley" (MC Rodolfinho).*

**SUMMARY**

The musical style that in Brazil is called Funk Carioca originated in the United States, but when it arrived in Brazil it acquired its own characteristics. The genesis of funk in Brazil over almost three decades led to the emergence of new sounds and themes, which, together with a favourable economic context and a language that cultivates ostentation, led to the emergence of the genre known as Funk Ostentação. This research looks at funk from its origins in the United States to the present day. An Ostentatious Funk scene has emerged in Presidente Prudente, which will be described according to interviews and fieldwork carried out with MCs, DJs and funkeiros; in the meantime, the reader will find the phenomenon analysed from a geographical and historical perspective.

KEYWORDS: Funk, Funk Ostentação, Territory, Technology, Consumption.

# CHAPTER 1 INTRODUCTION

The musical style known as Funk Carioca has been present in Brazil since the 1980s. Over the years, it has undergone changes in terms of the subject matter of the lyrics and also in terms of the sound and has given rise to a style, which emerged around 2008, called Funk Ostentação and which has been very successful among young people in urban areas. An Ostentatious Funk scene has also emerged in the city of Presidente Prudente; although limited and situated in the lower circuit of the urban economy, this scene exists and this work aims to analyse it. In order to do this, it is necessary to understand the genesis of funk and its developments, up to and including the phenomenon of Ostentatious Funk MCs in Presidente Prudente.

The idea for this research came about when I was teaching geography at a state school in Presidente Prudente and realised how interested many of the students were in funk and the culture of ostentation that accompanies Funk Ostentação. Since I needed to develop a research project to complete my bachelor's degree in Geography, I thought it would be a pertinent topic to work on, since it is a current issue in the context of youth phenomena and there is no bibliography - specifically on the phenomenon in Presidente Prudente. Initially, I searched social networks to find someone who had more knowledge on the subject and ended up learning about Tropa do DJ Robinho, which is made up of eight MCs. I spent a few weeks in contact with the Tropa and, from this contact, I was able to make the observations that enabled me to write this piece.

The final work has been organised into three chapters. In the first, the reader will find a historical reconstruction of the processes that gave rise to Funk, starting in the rural United States in the 1940s, going through its transformations decade after decade, until it landed in Brazil in the 1980s and then until the emergence of the phenomenon and development of Funk Ostentação, from 2008 to the present day. We also contextualise the socio-economic processes that have taken place in Brazil over the last decade and how they have contributed to the emergence of both Funk Ostentação and the culture of ostentation.

The second chapter contains the theoretical basis that underpins the entire construction of this work, covering concepts such as territory, the technical scientific informational environment, identity and compensatory consumption. This chapter establishes the relationship between the basic bibliography and the object of study, providing the reader with a better theoretical basis for the subject.

Chapter three is dedicated to characterising the local funk scene in Presidente Prudente. This chapter presents the scene's organisational structure and introduces some of the scene's main figures, such as DJ Robinho and his Tropa. In it I describe my visit to two funk dances in Presidente Prudente and report on the interaction I had with the people interviewed at these dances, emphasising that all the names used are fictitious in order to preserve the identity of the participants.

The last part is the conclusion of this work, which contains the final considerations that I have built up over the course of the research, based on the observations, questionnaires and interviews carried out; combined with the reading of the material that makes up the bibliographical basis of this work.

# CHAPTER 2 METHODOLOGY

With regard to the methodology and steps of the research, it was necessary to address aspects such as: social exclusion, elitism, prejudice, consumption, compensatory consumption, aspirational consumption, consumerism and ostentation. These aspects encompass the specific object of study, which is the phenomenon of Funk Ostentação in Presidente Prudente, a phenomenon in movement and transformation due to the plurality of its subjects (GIL, 1987), based on the assumption that exacerbated consumerism and the cult of certain brands would have the effect of reinforcing the self-esteem of these young people.

Analysing official data was essential in order to arrive at an answer regarding consumer power and market segmentation, which often deviates from the manufacturers' original proposal regarding the target audience, given the dynamics of the market. The data from the Brazilian Institute of Geography and Statistics (IBGE) relates to the consumption power and average income of class C in recent years; compared to the consumption power and income of classes A/B. In the 1960s and 1970s, the main discussion in studies about the middle class was about its political role and whether it was more aligned with the bourgeoisie or the proletariat, a paradigm that has changed in studies about the middle class in the 21st century, which have started to focus on consumption, largely due to the rise of a large number of people to the consumer market (BATISTELLA, 2014, p. 52).

In addition, interviews were carried out and a trial of insertion into the social group studied, along the lines of participant observation, which consists of analysing the object of study from within it, in the midst of those who make it up. Participant observation uses non-directive, non-standardised interviews, the main characteristics of which are: detailing issues or phenomena and the most precise possible formulation and organisation of related concepts; as little structure as possible, as well as the least possible interference by the interviewer during the interviewee's speech (COLOGNESE; MÉLO, 1998, p. 144).

As for the organisation of the interview script, I used the contextual script, which is commonly used in qualitative analyses that seek answers to aspects of life history and the reconstruction of social processes.

Colognese and Mélo (1998, p. 143) state that an interview is a process of social interaction in which the person interviewing aims to obtain information about the interviewee. However, an interview is not a loose conversation with no direction. The aim is to collect data that makes it possible to interpret the object of the research. I carried out two interviews with DJ Robinho's Tropa, one at a *snack bar* in Presidente Prudente when five MCs were interviewed. Another interview took place at a dance where I accompanied the MCs, immersed in participant observation. Interviews were also carried out with patrons of two dances, totalling 39 people interviewed.

Afterwards, I categorised the answers to the questions, relating them to the basic bibliography, making it possible to make statements based on the combination of the bibliography and the interviews. There was also a group interview with a collective of MCs from Presidente Prudente, due to the fact that we didn't have an agenda to meet each one separately.

Questionnaires were also carried out, which were applied to a wider public - if contrasted with those interviewed - seeking quantitative data on personal preference for certain brands - signs - that have strong symbolism in their relationship with ostentation. These questionnaires were multiple choice, with pre-set alternatives in line with the bibliography.

The so-called dichotomous modality was used in other questions that did not involve signs, but rather in more broadly polarising questions (CHAGAS, 2000). These questionnaires were administered to fifty young people aged between 14 and 18 living in Presidente Prudente in November 2015

The relationship between networks of influence and cultural pasteurisation, as well as the prevalence of the information environment, in the genesis and proliferation of the culture of ostentation were also addressed in the questionnaire, as well as the cult of exacerbated consumption from the perspective of Rucker and Galynsky (2008), who address compensatory consumption (ABDALLA, 2014, p.29).

# CHAPTER 3 THE HISTORY OF FUNK

Funk originated in the United States in the 1940s, when a large part of the black American population from the southern states moved to the industrial urban centres in the north of the country. Blues was brought with it, which took on electric characteristics, giving rise to Rhythm and Blues, a musical style that formed the basis of Rock n' Roll and went on to dominate the Billboard charts of the time, attracting even white artists who absorbed the black visual and sound aesthetic, such as the king of rock, Elvis Presley (VIANNA, 1987, p. 44).

Following the development of the sound, many artists became interested in new sounds that mixed Rhythm and Blues with the gospel music sung in black churches, and this new style was called Soul.

During this period, in the 1960s, Soul gave a voice to the black segment of American society in a historical context of struggles for civil rights and anti-segregationism, since laws that established separations between whites and blacks prevailed in a considerable part of the country, especially in the South. The main artists of this genre, such as Otis Redding and Sam Cooke, became an active voice for blacks alongside the Reverend Martin Luther King (Figure 1); tragically, all three died early in that decade, Otis Redding in a plane crash (still poorly explained) in 1967, and Sam Cooke and Luther King, assassinated in 1964 and 1968 respectively (KNUTSE, 2007; NEAL, 2013; EGUENDEL, 2014).

By the end of that decade, Soul was already being considered a retro style; Soul was considered a generic term for black music. Aiming to create new sounds, Funk emerged, characterised by more marked beats, with a lot of emphasis on the bass sound, heavier and more danceable. The pioneers of this style were George Clinton and his group Parliament, the group Earth, Wind & Fire, as well as Sly Stone and his group (Figure 2) (HERSCHMANN, 1997, p. 43).

Figure 1: Otis Redding, Martin Luther King and Sam Cooke Source: Rolling Stone; MTV

Figure 2: Sly Stone, George Clinton and Earth, Wind & Fire Source: Rolling Stone

Becoming a genre with high commercial acceptance, Funk became a favourite in clubs and nightclubs in the United States and, in the mid-1970s, underwent a change and began to have more pop-oriented beats, giving rise to the Disco style, which took over the world charts, with female names such as Gloria Gaynor and Donna Summer among its greatest exponents, as well as pop groups that adopted the style, such as ABBA and Village People, and of course the Bee Gees (Figure 3), who also adopted the Disco aesthetic. This period - between 1977 and 1981 - is considered by many to be the heyday of nightlife in the southern zone of Rio de Janeiro. It was a pre-AIDS era, permeated by the imminent return of democracy, which added to the festive atmosphere, and even Rede Globo invested in the industry, airing the soap opera Dancin' Days, one of the biggest hits of all time and which gave rise to the group As Frenéticas, which also became very popular (BAHIANA, 2006, p. 316).

According to Ana Maria Bahiana (2006), the soap opera Dancin'Days, which aired between August 1978 and January 1979, was a "happy union of narrative talents,

merchandising and opportunism that created "THE" soap opera of the end of the decade". Bahiana also states that the soap opera was the most perfect translation of the spirit of that time and that thanks to the success of the television plot, discos became a national craze, also taking a ride on the popularity of the film Saturday Night Fever (TURRA NETO, 2008, p. 143).

Figure 3: Clockwise: Bee Gees, Donna Summer, Village People and ABBA Source: Rolling Stone; TIME

In New York's Brooklin district - at the time, a stronghold of blacks and poor immigrants, very different from the gentrified Brooklin of today - and also in the Bronx district, on a smaller scale, the Sound Systems phenomenon emerged and gained momentum. Sound Systems originated at the end of the 1960s with Kool Herc, who exported the *block* party model from Kingston, Jamaica. At these open-air parties, black music LPs were played (Soul, Rhythm and Blues, with *scratch* techniques[1] , which produced a peculiar sound. Sound systems only gained more prominence in the early 1980s, with artists who are considered disciples of Kool Herc, such as Afrikaa Bambaataa and Grandmaster Flash (Figure 4), precursors of the Hip Hop movement. While these DJs played with *scratch* techniques *on* funk records, they improvised on the microphone and dancers danced to the music, these were the DJ and the b-boys, respectively (VIANNA, 1987, p.43).

This so-called hip-hop style soon became popular with groups such as Run DMC and the Beastie Boys, who were probably the first hip-hop group made up of white people, and it should also be noted that the members of the group were Jewish (VIANNA, 1987, p. 49).

Figure 4: Kool Herc, Afrika Bambaataa and Grandmaster Flash. Source: MTV

At the same time, in Miami, Florida, a new sound was beginning to emerge which, although it didn't spread into the *mainstream*, ended up being the root of other sounds, including funk carioca. This new sound became known as Miami-Bass and was characterised by drums and chimbais in a start-stop rhythm. Its main exponents were MC ADE, Maggotron and the 2 Live Crew (Figure 5) (FUNK RIO, 1994; STYLUS, 2005).

In Brazil, since the 1970s in Rio de Janeiro, there have been dances that played the main Funk and Soul hits. These dances were mainly held in the northern suburbs and had their organisational basis in the Equipes - which can be compared to the Sound Systems of New York. These teams promoted several dances every weekend - the biggest being Furacão 2000 - at a time when spontaneity and ease of organising an event were not yet a reality, largely due to the difficult access to Funk and Soul LPs, which were imported from the United States and cost a lot. The transition from these suburban dances to Funk Carioca dances was gradual. Given the difficulty of accessing various imported discs, some DJs started using Miami-bass discs, which had their bases isolated, giving rise to funk carioca. The beat became popular and was used by most DJs at dances in the periphery (VIANNA, 1987, p. 94; MEDEIROS, 2006, p. 16).

Figure 5: MC ADE on the Xuxa programme in 1994; 2 Live Crew Source: Globo; Hip Hop Nation

The Miami-bass beats that would become the characteristic beat of funk carioca were produced on a synthesiser called the Roland TR-808, which cost just over US$1,000, a steep price for the time, but still one of the most affordable synthesisers on the market.

At this time, towards the end of the 1980s, the first funk artists from Rio were released through compilations, some radio stations started playing the sound and its popularity spread to the slums and favelas. The public in the south of Rio de Janeiro had until then paid much more attention to the national rock style of the 1980s - considered to be the golden decade of national rock - and funk, although gaining popularity, prevailed only in clubs and nightclubs on the outskirts, while it gradually began to be introduced into the set lists of clubs and nightclubs in the south, also gaining popularity with this more affluent public (HERSCHMANN, 1997, p.81).

The great promoter of the style at that time was DJ Marlboro (Figure 6), responsible for the Funk Brasil compilation of 1989, which featured various artists. The main hit was Feira de Acari, by MC Batata (Figure 6). The song is about the Acari Fair, in the suburb of the same name in Rio de Janeiro. This fair is famous for selling products far below the normal selling price, supposedly because they come from robberies and assaults (STAMBOROSKY, 2009).

Figure 6: Cover of the 1989 Funk Brasil compilation; DJ Marlboro and MC Batata Source: R7

As the 1990s began, compilations continued to be released and funk began to reach an ever larger audience, both inside and outside the favelas, and consequently received more media attention. As had happened with Samba in its origins at the beginning of the century, funk began to be stigmatised by the mainstream media. A characteristic of the funkeiros at that time was that they left their favelas and slums to go to the dances and beaches in large groups, the so-called galeras. There was rivalry between the galeras from different favelas and sometimes there were episodes of violence, both at dances and on the beaches of the south zone. Instantly, funk became a public enemy, with recurrent newspaper and television reports portraying funkeiros in general as thugs and delinquents, largely due to confrontations between rival galeras, which were reported by the media as "arrastões" in the southern zone of Rio de Janeiro in 1992 (HERSCHMANN, 1997, p. 15).

> He said that at the market for the price of a jerrycan
> I used to buy the fridge, the pans and the cooker, All this you
> can find, in a street just over the road, it's easy to find, It's
> there at the Acari market (MC Batata)

The negative media visibility drew attention to funk and the rhythm began to be played outside Rio de Janeiro. In this context, funk carioca began to attract the attention of the big music conglomerates, who saw an opportunity to capitalise.

In 1996, the EMI label released Claudinho & Buchecha's first album (Figure 7), which bore the duo's name. The album reached the top of the national charts and sold over a million copies, propelling the former bricklayers to stardom with songs such as Ra do Sal ueiro and Conuista. The stylishly produced album launched a new genre called funk-melody, a more pop-sounding funk that was accessible to the general public, focussing on issues related to relationships, leaving aspects such as violence and social problems (STAMBOROSKY, 2009) in the background.

Figure 7: Claudinho and Buchecha Source: Midas Studio

The Furacão 2000 team began airing a weekly programme throughout the country and, although it didn't make funk a national phenomenon, it took it to previously unexplored places. The rhythm also began to be played freely on radio stations in Rio, without the restrictions of the beginning of the decade, largely due to the more accessible style of melody funk.

> I want to find you, I want to love you

You are everything to me, my land, my sky, my sea
I want to find you, I want to love you (Claudinho and
Buchecha)

At the turn of the century, funk underwent a new stylistic reformulation. Prohibidão funk emerged, given its name because it deals with themes related to sex, which is portrayed explicitly, as well as exalting criminal groups and factions, such as the Comando Vermelho. At the same time, funk continues to be in evidence, either through the "proibidões", which are produced amateurishly and are not played on radio stations due to the apology they make, or through artists more closely linked to traditional Rio funk, such as Bonde do Tigrão (STAMBOROSKY, 2009, SALLES, 2014).

The first decade of the 2000s saw funk carioca gain international exposure and be recognised as a unique phenomenon. It received articles in various international media outlets and was incorporated into the repertoire of non-Brazilian artists, such as the British M.I.A., acclaimed for her 2005 album Arular.

Tati Quebra-Barraco (Figure 8), who became popular with her songs full of ambiguities and sexual connotations, was even the subject of a documentary about the new wave of female MCs (SOU FEIA MAS TÔ NA MODA, 2005).

Naturally, funk has always dealt with themes that are pertinent and everyday for funkeiros - artists and audiences alike. As such, violence and sex have always been hot topics. It just so happens that funk, already popularised throughout Brazil, has established its own scenes in São Paulo and also in the Baixada Santista. The socio-economic rise of a large number of people from the less favoured classes into the consumer market since 2003 led to the emergence of a new style around 2008: Funk Ostentação. Just as the blues electrified when it left the South of the United States and made its way to the Northern states, funk from Rio - full of social laments and apologies for violence - acquired a cult of consumption guise as the local scenes in São Paulo and the Baixada Santista developed, associated with a favourable economic context.

However, even before this new version of funk, there were isolated cases of ostentation-themed songs, such as MC Orelha's hit "Gaza Strip" (Figure 8), which is a mixture of prohibition and ostentation.

In the Gaza Strip it's all bombs, in war it's all or nothing, several titaniums in the comb and bulletproof vests. We go down to the track to do the robbery but we're locked in the 12, if I'm on a roll it's 600 bolados, imported perfume, pistol in the holster. We don't need credit, we pay for everything in cash. It's Ecko, Lacoste, Oakley and various team shirts. Outsiders even think it's easy to live off crime (MC ORELHA).

Figure 8: Tati Quebra Barraco and MC Orelha
Source: Extra; GR6

## Ostentatious Funk

First of all, I would like to warn the reader that as this is a recent phenomenon, there are not many theoretical references on the subject of study. As such, the history and developments that have made Funk Ostentação a widespread national phenomenon will be presented on the basis of what has been reported in the media over the last few years. Funk Ostentação is an offshoot of the already traditional Funk Carioca and emerged around 2008. The first funks of this style emerged in the city of São Paulo, more precisely in the district of Cidade Tiradentes, in the far east of the city, and in the Baixada Santista, a region where funk has historically been more present in the state of São Paulo. The MCs Backdi and Bio G3 (Figure 9) probably released the first song in the genre, called Bonde da Juju, in September 2008. Bonde da Juju was a clear reference to expensive objects, such as Oakley Juliet glasses. At festivals and, little by little, at concerts in São Paulo and the Baixada Santista, MCs of the genre began to emerge, all of them addressing in their lyrics the consumption of high-value products such as cars, motorbikes, drinks and, above all, clothes; as well as mentions of social ascension and life outside the confines of the favela, with everyday things such as going to the beach or the shopping centre being considered ostentatious situations. In 2011, Funk Ostentação began to gain national notoriety, largely due to Megane, by Boy do Charmes, who recorded a video clip to promote the song, setting the trend for a video aesthetic that became associated with the genre. Professional videos have never been commonplace in the funk world, with the exception of veterans Claudinho and Buchecha in the mid-1990s (PEREIRA 2014).

> You're wearing Juliet, Romeo 2 and Double Shox
> 18 K on the neck, from Ecko and Nike Shox
> You're wearing Juliet, Romeo 2 and Double Shox
> Worth more than a baron, this is the Oakley tram (MC Backdi and MC Bio G3)

According to the documentary Funk Ostentação - O Filme, in an attempt to leverage the media power of Funk Ostentação, the production company KondZilla emerged, specialising in producing clips of the genre and achieving great success after the release of Megane. The success was such that other producers started investing in the same genre and practically all emerging MCs or those with higher expectations of a particular song started recording ostentatious clips, full of booze, cars, motorbikes, jet skis, mansions, jewellery and expensive clothes, as well as lots of women.

I'm Daleste with the Angra tops next door
São Paulo is ostentatious, his is tin, mine is gold (MC Daleste)

Konrad Dantas, 25, is Kondzilla, and he says that his inspiration for recording the clips is the aesthetics of American *gangsta* rap, with various references to ostentation. Kondzilla has all his clips signed, i.e. with his name at the beginning and end. All the clips he has been involved with have already passed the 100 million views mark on YouTube. At the same time, other production companies specialising in funk have emerged, such as Funk TV, based in Cidade Tiradentes, in the eastern zone of São Paulo (PEREIRA, 2014).

According to Pereira (2014, no pagination):

> In the video clips they record and upload to *YouTube*, they present themselves as rich characters (or as they imagine what it would be like to be rich), who amuse themselves by showing off expensive products and large sums of money, sometimes throwing them up in the air. In a way, they're trying to turn imagination into reality.

In 2012 and 2013, three out of the ten most watched videos on YouTube in Brazil were Funk Ostentação clips. MCs such as Rodolfinho, Daleste, Lon and Guimê (Figure 10) have become real celebrities, performing all over the country and inspiring many young people to try a career in funk. The media began to give space to news related to funk ostentação not only as a musical style, but also as a social phenomenon (FUNK OSTENTAÇÃO - O FILME, 2012; PINHEIRO-MACHADO E SCALCO, 2014).

> Imagine me in a Megane, or a 1,100 invading the proms, there won't be anyone left Our tram as it goes It's euros, dollars and 100s.
> Note of 100, note of 100 (MC Boy do Charmes)

In July 2013, MC Daleste, one of the most prominent of the genre, was murdered during a concert in a COHAB in Campinas - SP. To this day, the crime remains unsolved and has caused a great deal of commotion among fans. Daleste's death was widely reported in the national and international media and the term "Daleste" was one of the 10 most searched terms on Google in Brazil that year. Media interest soared and the genre of Funk Ostentação gained even more prominence in the media (TERRA, 2014).

Figure 9: MC Backdi and MC Bio G3
Source: R7

The rise of Funk Ostentação is inseparably linked to the emergence of Brazil's so-called "new middle class" since 2003. State programmes for access to income and housing have led to around 42 million people leaving the D/E classes to join the C class. The enormous growth of class C has led to an increase in consumption in general by these people, and has also favoured the consumption of ostentatious products, which are now less difficult to access. The most desired products are cars and motorbikes, but the most accessible and characteristic of the culture of ostentation are those related to clothing - to visual identity - such as trainers, glasses, watches and clothing from brands with status within this culture (KREPP, 2014).

The 1980s and 1990s were marked by moments of insecurity regarding the solidity of the national economy. The period was permeated by striking situations such as rampant inflation in the 1980s - dubbed by economists as the lost decade[2] - the confiscation perpetrated by the Collor Plan[3] , as well as the difficulties faced while the Real established itself as the national currency in the mid-1990s. The moment of stabilisation and economic growth concomitant with a significant reduction in poverty levels only came in the following decade, from 2003 onwards (MOREIRA, 2011).

The unequal distribution of income has been a historical problem in Brazil, and this issue is constantly being addressed. Indexes specialised in measuring income concentration, such as the Gini[4] , rank Brazil among the societies with the highest levels of income concentration. In this way, social inequality is a topic that is constantly addressed when it comes to the country's development, not only in terms of economic aspects, but also in terms of justice and the social affirmation of the less favoured classes (DAYRELL, 2001, p.106).

---

[2] The 1980s is considered the lost decade by economists due to crises and little economic progress, permeated by galloping inflation in Latin American countries.

[3] An economic plan put into practice by the government of Fernando Collor de Mello, which took controversial measures, including a limit on people's ability to withdraw from their current accounts.

[4] The Gini coefficient or index is a measure of inequality developed by Conrad Gini and adopted by the EUROPEAN COMMISSION.

Figure 10: Clockwise: MC's Rodolfinho, Lon, Boy do Charmes and Daleste Source: R7; FOLHA DE SÃO PAULO

Over the last decade, class C, known in the media as the "new middle class", has grown significantly in Brazil. According to Neri (2010), 94.9 million people were part of the Brazilian middle class in 2009, which represented around 50% of the Brazilian population at the time. In 2012, this percentage would comprise 53 per cent of the population, according to an estimate by the Secretariat for Strategic Affairs of the Presidency of the Republic (SAE/PR) (2012). The growth of the middle class is the result of a large number of people leaving classes D and E and joining class C. This phenomenon is due to a reduction in social inequality and better income distribution, as evidenced by the Gini index, which has been falling steadily over the last decade (BATISTELLA, 2014 p. 48).

The rise to class C was quite substantial, according to the Ministry of Social Development (MDS) (2011), over the previous eight years; the number of Brazilians who left extreme poverty and entered the consumer market was approximately 22 million. Government income distribution policies, such as the Bolsa Família (Family Allowance) programme and access to mortgage credit through the Minha Casa, Minha Vida (My House, My Life) programme, have been responsible for guaranteeing the basic aspirations of many people who belong to the low-income population, such as access to income and housing.

The favourable times, supported by easy access to credit, brought these people into the consumer market, while the heads of household began to aspire to consume products that were previously inaccessible. In 2014, according to SERASA (2014), the favourite products were plane journeys, washing machines, plasma, LCD and LED TVs, as well as, of course, cars and houses. Alongside the new aspirations of the heads of household, there are also the consumer aspirations of their children and/or grandchildren. These young people, born and raised in a privileged economic context and with greater social equality than that faced by their parents, find themselves more concretely inserted into contemporary society through broad access to information, especially through social networks (BATISTELLA, 2014).

The level of illiteracy in Brazil has fallen considerably in recent decades. In 1992, illiterate people accounted for 17.2 per cent of the Brazilian population, while in 2012 they represented 8.7 per cent of the population (IBGE, 2012).

One of the conditions for receiving Bolsa Família benefits is to keep school-age children in school, and although programmes to promote access to higher education have taken a large number of young people to university, through FIES and PROUNI, education and employment are still a problem for this generation, given that the percentage of "nem-nem" is considerable, i.e. although they are better educated than their parents, young people still don't have full access to education and/or work. The term "nem-nem" refers to those aged between 15 and 29 who do not study or have a paid job. In 2014, according to the IBGE (2014), 20.3 per cent of young Brazilians in this age group were in this situation, and in the Southeast, the country's most economically developed region, this rate was 37.9 per cent, the highest rate among the country's regions.

Concomitant with the arrival in adolescence of young people who grew up in this favourable economic and social context - especially those born after 1997 - new social representations emerged, manifested through the "culture of ostentation" and "rolezinhos". From this perspective, clothes, shoes and accessories become more than just objects, they are characterised as social signs that relate to aspects of self-affirmation and compensatory consumption (ABDALLA, 2013, p.27). The objects of desire for these young people are now high-value products, traditionally linked to the A/B classes.

Ostentatious funk emerged in the peripheral regions of the Baixada Santista and the city of São Paulo, where a significant number of people who are part of the new C class and who have recently entered the consumer market live.

This type of funk seems to express the aspirations of a new generation that aspires to consumption, combined with aspects of pure consumerism, since basic needs such as housing, school, sanitation - the aspirations of their parents and grandparents - would already be met due to more concrete public investment, as well as the considerable increase in the average income of citizens in the previous decade (BATISTELLA, 2014 p. 46).

This new generation, with ample access to technology such as computers and mobile phones with Internet access, has a great capacity for articulation, since the phenomenon of ostentatious funk happened spontaneously, since the genre's productions are not promoted by the mainstream media or phonographic conglomerates, but rather by independent producers, when the productions are not made in home studios (PEREIRA, 2014).

According to McLuhan (1964) *"the medium is the message"*, this famous phrase aims to explain how technology - especially information technology - has the ability to influence and mould aspects of the society in which we live. Around 2013, another phenomenon emerged on the Internet, more precisely on social networks: the "rolezinhos".

For a long time, shopping centres were "fantasy islands" where the wealthy could move around safely in an air-conditioned environment while they shopped or had a snack. In shopping centres, the prices of products and services are generally higher than in similar establishments on the street (PADILHA, 2006, p.14). From this perspective, shopping centres have traditionally been the domain of the A/B classes. However, the aforementioned favourable economic context, combined with the consumer aspirations of young people from generation Z, has led them to frequent shopping centres, giving rise to the phenomenon of "rolezinhos". The "rolezinhos" are gatherings organised through social networks, or information and communication platforms such as Facebook and WhatsApp. Attendees - mostly minors - arrange to meet in shopping centres, arriving in their hundreds and occupying the space.

Social networks are also fertile ground for articulating these phenomena, disseminating information at exponential speed. According to Wasserman and Faust (1994), quoted in Recuero (2009), social networks are a combination of two distinct elements, actors and connections, each with its own role. Actors are people, groups and/or organisations, while connections are social ties and relationships, interactions, contact. Among the new dynamics brought about by the advance of the information environment, specifically social networks, is the fact that anyone can socialise with others and/or express themselves, wherever they are and whoever they are.

In this context, still with ostentation as a backdrop, there is another characteristic phenomenon: the famosinhos. The famosinhos are young people from generation Z, aged between 13 and 20, of both sexes and living in peripheral areas, usually in the C class. They maintain official Facebook pages where they post photos and videos showing off designer clothes. These pages are liked by thousands of fans. The famosinhos inevitably dress in a similar way to the funk MCs, and are true celebrities in the *undergroud*, although they are famous simply for being famous. The famosinhos have their existence based on social networks, being linked to ostentation and the prevalence of the information medium. Even the rolezinhos, before they were known as such, were just gatherings of famous people and their fans. These gatherings could bring together several hundred fans. The "idols" - famous for being famous - sometimes left these gatherings covered in gifts, some fans spending as much as R$100 on presents (KREPP, 2014).

The attempt to "include" young people from class C and the periphery, however, was met with resistance. In some "rolezinhos", there were incidents of vandalism by young people, which generated a great deal of debate in the media about the phenomenon, almost always being approached in a negative way, being reduced to a simple mess and disorder and not seen as a social phenomenon, which may be a consequence of the entire basic structure of Brazilian society. Although they were there in a legitimate way, they were repressed by the shopping centres, which, in a way, "choose" who they want their public to be, associating their image with that target public (FIGUEIREDO; CAVEDON; SILVA, 2013). Some shopping centres have filed lawsuits to have the courts officially prevent unaccompanied minors from entering the malls - trying to avoid this C-class and peripheral public. Court injunctions guaranteed what the shopping centres had proposed. In other "rolezinhos", there was rioting and running around, with clashes between police and minors, who ended up being apprehended, without justification (KREPP, 2014).

The rejection of these young people also comes from the A/B consumer public, who have started to reject the products they used to consume, under the pretext that they have become "poor people's stuff", in other words, the value of the product would not be in its usefulness and quality, but in the exclusivity and status that its possession would bring. Some brands that are among the favourites of Generation Z youngsters have indicated that their products are aimed at the A/B classes and consulted Data Popular (2014) to see what negative impact associating them with the C class public would have on the brand, as they are considered poor.

Although they don't usually take this stance publicly, some expensive brands feel ashamed of these consumers (KREPP, 2014). Ironic is the fact that, according to data collected by Data Popular (2014) and published in January 2014, young people in class C have a higher total income than those in classes A/B and D combined. While young people in class C have an income of around R$130 billion, the other classes mentioned barely reach R$100 billion (Data Popular, 2014). In 2009, class C was responsible for 46.24 per cent of the Brazilian population's purchasing power, being the dominant class when it comes to consumption, yet they are still rejected or relegated to the background in many situations. In this sense, it's important to note that the terms used by the mainstream media to make descriptions don't always correspond to scientific concepts. Take, for example, the definition of social class. As for social class, it can't be defined solely on the basis of income level, but there are other issues such as habits, lifestyle, education and culture. According to Jessé de Souza, from the Centre for Research on Social Inequality (CEPEDES) at the Federal University of Juiz de Fora:

> The "real" middle class in Brazil is no more than 15% of the population. They are the people of "privilege" (...) which is 'cultural capital', we're talking here about 'valued' cultural capital, that technical or literary knowledge that guarantees prestige and high salaries. Not only does the differentiation of cultural capital help to create distinct classes, but cultural capital is also constantly used as a mechanism of social distinction. Thus, 'noble' taste legitimises those who possess it to despise classes with supposedly vulgar tastes. This kind of prejudice and solidarity based on taste is a modern form of class consciousness and solidarity that helps to reproduce all kinds of unjust privilege (SOUZA, 2013, unpaginated).

In an article published by the R7 portal, Vera Batista, professor of criminology at the State University of Rio de Janeiro (UERJ) and general secretary of the Carioca Institute of Criminology, says that Afro-Brazilian cultural expressions in our history have always been criminalised, as has capoeira and samba:"It is already a tradition to look at the cultural expressions of the poor, especially Afro-Brazilians, with this superior and prejudiced gaze"; and it is possible to associate such aspects with the concepts coming from the classic work, Roots of Brazil (HOLLANDA, 1936), which addresses Brazilian society as racist, Eurocentric and elitist; aspects masked by the legend of the "cordial man" and which are still current even after almost a decade.

# CHAPTER 4 TERRITORY, ENVIRONMENT, TECHNOLOGY AND CONSUMPTION

**Territory**

In its etymology, territory comes from the Latin "territorium", and for a long time was present in treatises on surveying, having the meaning of an appropriate piece of land (HAESBAERT, 2004 p. 43). When analysing the cultural phenomenon of funk ostentação geographically, the geographical concept of territory is very valid. According to Raffestin (1993), territory goes beyond geological and geomorphological concepts, since territory is the expression of relations of power and domination, appropriated in a given space. The traditional political territory is the space administered by a greater power - in this case the state - which establishes its guidelines and management, making its power present over those who live within that space delimited by borders.

Territory, however, can be characterised in another way, through multi-scalarities of power and, consequently, multiple territorialities, inserted in the same context of geographical location. Take, for example, any favela located on a hill in the south of Rio de Janeiro. The favela in question is part of the state's political territory - at least in theory - as it is geographically located within the borders of the municipality, state and country. At the same time, however, based on the supposed frequent activity of drug trafficking (which takes place freely within the confines of the favela and the absence of the state through its obligatory actions, such as policing, for example), the favela becomes the territory of drug traffickers, who exercise their power within that area. It is thus the territory of the state and, at the same time, the territory of the drug traffickers. This scalarity can also be exemplified by citing the fact that the majority of favela residents are black, brown and poor, which makes the favela their territory: black/brown and poor, with those who are, for example, white and rich not necessarily suffering discrimination, but being considered as individuals outside their territory (SOUZA, 2001, p. 77).

> If classic territorialities - including that of the nation-state, as conceived within what Foucault calls "sovereign power" - indicated a clearer distinction between inside and outside, the same and the different, the "native" and the foreigner, it should finally be emphasised that we are now dealing with a territorial dis-order marked much more by ambiguity, where the very process we have constructed to "contain" the other actually "contains" us and where the "other" - as in the case of the great South-North migratory flows - is increasingly in "our" territory, on our side (HAESBAERT, 2009, p. 118). 118).

Dematteis (2008), in turn, states that territory is a social product, a place of relationships, taking into account the interactions that occur between different places and people; it is a social construction, where there are inequalities associated with the natural and specific characteristics of each place. It is the result of a collective construction with different dimensions, taking into account multi-scalar relations, changes, conflicts, domination and social fabrics (DEMATTEIS, 2008, p. 34).

In this way, it is possible to understand that a territory cannot be delimited taking into account only the landscape, because territory is not always a physically constructed entity, even though its materiality is undeniable and influences its own constitution. In order to delimit a territory, the social subjects that are territorialised have to be taken into account. The forms taken through social relations of power, projected onto space, are quite diverse (TURRA NETO, 2008, p. 466).

In the same vein, Turra Neto (2008), based on Haesbaert (2004), states that territory is defined on the basis of social relations and the historical and geographical context in which it is located. In this way, territory is relational in two ways: firstly, because it is inserted within the framework of historical and social relations and because it includes a complex relationship between social processes and material form.

> For a long time, the traditional idea of territory was associated with a contiguous area, delimited by a homogenous culture or by a nation state, with sovereignty and control of its borders. Traditional geography and political science were largely responsible for establishing this perspective, which is usually implicit in contemporary deterritorialisation discourses, whether they are more political, economic or cultural (TURRA NETO, 2008).

Territories don't necessarily have to be continuous, they are dynamic and their area of influence can encompass other territories, but they can also be de-territorialised, i.e. leave behind characteristics associated with a particular territory. Discontinuous territories are juxtaposed and constructed in movement, and it is possible for an individual or group to be deterritorialised in some aspects and, at the same time, reterritorialised in another aspect. In this sense, there are network territories, which are configured according to the logic of networks, being discontinuous and susceptible to being superimposed; the control of space taking place through the control of flows and/or connections, network territories are in tune with the multiplication of territories, as well as the experience of a multi-territoriality (TURRA NETO, 2008, p. 469).

Just as Samba originated in the slums - which was its territory - and gradually began to expand outwards, first within the city and then in the national context, until it reached the international stage, being recognised as a symbol of the country worldwide; Funk Carioca went through a similar process. Initially, the territory of funk was precisely the favelas and suburbs where a large number of fans of the style lived. Appropriation by the media and *mainstream* record labels expanded the territory of funk beyond the borders of the favelas, but funkeiros caused strangeness when they frequented beaches in the south zone, and this space was appropriated under the guidelines and power of those who live there, i.e. the middle and upper classes. So, although the territory of funk had expanded, the territory of the funkeiro had not yet, although it is not a static and permanent definition, due to the cyclicality and dynamics of time scales. In relation to these aspects, Souza (1995, p.81) states that:

> Territories exist and are constructed (and deconstructed) on the most diverse scales, from the smallest (e.g. a street) to the international (e.g. the area formed by all the territories of the member countries of the North Atlantic Treaty Organisation - NATO); territories are constructed (and deconstructed) on the most diverse time scales: centuries, decades, years, months or days; territories can be permanent,

but they can also have a periodic, cyclical existence.

According to Magnani (1996), young people have a latent need to be among equals in places that are defined by their "exclusive marks". These are the places where they can "exercise in the use of common codes, appreciate the symbols chosen to mark differences. This is how the network of sociability is built (apud TURRA NETO, 2000, p. 96).

Since music is an immaterial phenomenon, it is difficult to establish limits for its manifestations, based on the fact that music is a universal language and its territorial scale can expand or recede according to social and economic contexts and also due to fads, which are becoming very frequent at a time of widespread access to information, as a result of the technical-scientific-informational revolution.

**Technical Scientific Informational Environment**

One of Brazil's most celebrated geographers, Milton Santos, defined a concept that is very widespread and problematised in the study and analysis of the human sciences, the concept of the technical scientific informational environment. The term covers the development of the dynamics of production and reproduction in the geographical context. Space is constantly transformed dynamically - slowly or more quickly, depending on the specific case - from the natural environment to the technical and informational environment (SANTOS, 1996 p. 156).

The **natural environment** is defined as the period in which the production of space was dependent on nature as an omnipresent delimiter, with anthropic actions being less invasive and the legacy left being less significant. The natural environment can be defined as the initial period in the process of production through anthropic activities. The peoples who lived during the long period in which organised societies originated were very dependent on the natural environment and the activities they carried out had a much smaller impact than the anthropogenic activities carried out today. This shows that human actions had clear limitations in terms of their legacy and that the ability to restore nature naturally was at a higher level.

The **technical environment**, in turn, symbolises mechanisation and the adoption of technologies that alter the archaic dynamics of production, creating growth and revolution, with the First Industrial Revolution being a milestone in this sense. The International Division of Labour and the need for anthropic actions, in terms of the management of new technological instruments and machinery, in favour of production, are also factors listed in this analysis. The passage of time has improved human knowledge in relation to specific instruments and techniques, which have enabled man, to a certain extent, to interfere in the dynamics of nature, expanding his capacity to produce space, which has become a technical and instrumentalised space. With regard to these instruments, Milton Santos (1996, p. 158) stated that: "they are no longer extensions of your body, but represent extensions of the territory, true prostheses"

The **technical scientific informational environment** thus symbolises the current times in which we live, in which the globalised capitalist system leads to the production and transformation of geographical space through the combination of the factors: Information and Technology. The Third Industrial Revolution has acquired characteristics that fit into this context, having been accentuated since the 1970s and its impact having become evident in such a way that this process has been called the Informational Scientific Revolution. The advance of technological discoveries and the adoption of techniques and instruments for production have been present all over the world, characterising the process of globalisation in the new construction of geographical space and also in the way individuals tend to deal with it. Information has become the driving force behind globalisation. The democratisation of technologies promotes homogeneity up to a certain point, since the process is not consolidated in the same way everywhere, only adopting related aspects without effectively inserting itself into the globalised world in various places (SANTOS, 1996, p. 27). In this way, it is completely possible for an MC in Presidente Prudente to record his songs and make them available on the Internet for free access; however, making a professional quality recording at the level of studios where major artists record is quite unlikely, given the technical limitations of their equipment.

**The Technical Scientific Information Revolution or Third Industrial Revolution**

The phenomenon known as the Technical Scientific Information Revolution, or even the Third Industrial Revolution, came into force from the late 1970s onwards, based on a large number of technological developments, mainly related to the dissemination of and access to information and knowledge, through information technology and telecommunications. Advances in areas such as chemistry, biochemistry, medicine, genetics and robotics were also present, but the revolution in communications, and consequently access to information and the dissemination of knowledge, was certainly the most prominent point, becoming part of the entire production chain, whether in space or not. The advance of the so-called globalised world has one of its main foundations in the technical-scientific information revolution, providing greater dynamism in the transit of goods, capital, people and, above all, information and diverse knowledge. Satellite TV and radios alone increased access to and dissemination of information, but it was the advent of the dial-up Internet that provided almost instantaneous access to an inexhaustible range of knowledge that was previously inaccessible to people who were not part of the geographical, social or political context of the source of the information in question.

The Internet combined with information technology makes it possible, for example, for funkeiros from the Funk Ostentação genre to make their productions amateurishly with few resources, making it available to a (possible) wide audience. This situation was impossible before the advent and improvement of information technology and the Internet, since popularisation depended on less democratic means of dissemination, such as radio and TV, which have commercial interests and more restricted agendas

**Identity**

The word identity appears in dictionaries as a combination of the Latin words *idem* (the same) and *entitas* (entity), meaning: "the same entity". The value of identifying people is a recurring theme in contemporary sociological approaches, whether it's identification as a member of a particular social class (ABDALLA, 2014, p. 25; ORIGEM

DA PALAVRA, 2014).

An individual can acquire various identities over a period of time, either momentarily or even permanently, according to the geographical and social context in which they find themselves (HALL, 2001, p. 54).

The way in which an individual defines themselves in relation to their own identity is called self-concept. Self-concept relates to the social group or groups to which the individual believes they belong or don't belong, making an assessment of themselves. Naturally, all groups have their own characteristics in terms of symbolism and rituals and/or customs. As an example: an individual who supports a certain football team will probably have a shirt from that team, even if it's a crude replica bought from a street vendor. Therefore, an individual who likes funk music will do their best to adjust their clothing to what is characteristic of the style's followers, and the characteristic clothing is the bearer of the symbolism that surrounds the identification in question. In this way, individuals affirm and express their identity through possessions, either individually or as part of a group.

According to Dayrell (2001), funk dances and some of their rituals have existed since the entry of Miami-bass into Brazil, but the act of assuming themselves as funkeiros, as an expression of the style they adhere to, as an identity, only became widespread from 1995 onwards. The look

would then be an element that, in some way, contributed to the funkeiro's identity, as MC Fred from Belo Horizonte - MG said in an interview with Dayrell: "People already programme a pair of shorts or trousers with the waist down, crazy trainers, a little hat, a T-shirt..."

The search to belong ends up making young people attracted to funk dress in a similar way to the MCs, within their means, with clothing being just another factor added to the funkeiro's identity. According to Dayrell (2001, p.221):

> Being a funkeiro doesn't imply a set of common values and behaviours, like a religion, but it does constitute a specific way of experiencing the demands of this phase of life. The identity of funk is offered by the style of possibilities for living and expressing the drives, desires and needs that characterise the youth condition. Thus, funk's commitment seems to be to joy and fun.

Individuals who are still defining their identity and their place in society adopt the consumption of products that can place them in a group with which they feel identified; children who have autonomy to define their clothing - usually after the age of 10 or 11 - tend to adopt clothing similar to that of their peers of the same age, or of adults they admire and have as idols. At this stage, it is common for a relationship to develop that is antagonistic to the desires and approval of parents, concomitant with a search for acceptance from other individuals in the same age group (HALL, 2001 p. 56).

The dynamic of the search for belonging and acceptance is linked to human sociability, since the human essence is first and foremost social. Wallon and Vygotsky assure us that man is genetically social and that the self and the other are forever linked in this contextualisation, in which the relationship with oneself presupposes the relationship with the other (DAYRELL 2001, p. 227). In this way, identity as a funkeiro means hope, direction and meaning in life, which is a lot for young people condemned to a life without meaning or hope (DAYRELL, 2001, p. 224).

During this period of adolescence, young people have many different cultural references, which make up a heterogeneous set of networks of meanings that are articulated and acquire meaning in their daily actions. In this way, young people can interpret their social position, giving meaning to the set of experiences they live through, being able to make choices by acting on their reality: the way they represent themselves as subjects is the result of these multiple processes they have gone through (DAYRELL, 2001, p. 234).

A person's identity even extends beyond their body, adding "me" and "mine" to what Belk (1988) calls the extended *self*, encompassing the body, psyche, possessions, assets, reputation, family relationships and friendships with others. "Having" and "being" are not the same thing, but they are inseparable, and it is understood that according to the proportion of the subject's relationship with an object, the greater the chance of the object in question becoming a good, i.e. something that the consumer is zealous for, preserves and enjoys, worrying about not destroying it since the object is important in their daily life. Through consumption, people express their identity, communicating non-verbally, letting consumption speak for itself (ABDALLA, 2014, p. 29).

**Compensatory consumption**

A large number of individuals choose to consume products that are characteristic of the identity they would like to have, since consumption itself is not only related to their self-concept, but also to the aspirations they may have, either in relation to a specific person they look up to as an example, or to a specific group they are not part of, but would like to be. Analysed from this point of view, the value of the product a person consumes is linked to the symbolism it presents, within the context of conveying an image of themselves that suggests a socio-economic reality that doesn't necessarily match the one they're actually in. In other words, the individual socially conveys the false impression of being successful to their peers and strangers through their clothes and accessories - especially mobile phones - even though they have a low income (ABDALLA, 2013, p. 26).

Glória Diógenes (1995) states that the expansion of industrialism, mass culture and consumerism are the keynote of a new era, in which youth is a catalysing agent and propagator of a cosmopolitan style with characteristics of modernity. Morin (1990) corroborates this, saying that youth is a fundamental element of modern culture. Young people are the most expressive segment in the consumption of symbolic goods that are produced and reproduced by mass culture. The message transmitted in this cultural context is that consumption is within everyone's reach, as well as the pleasure of consuming (HERSCHMANN, 1997, p.117).

> To talk about youth is to move in an ambiguous field of conceptualisation. Youth is constituted as a social category in terms of the definition of an interval between childhood and adulthood only at the end of the 19th century, gaining clearer contours at the beginning of the 20th century. Youth is a modern invention and is thus woven into a terrain of constant transformations (DIÓGENES, 1998, p. 93).

According to Rucker and Galinsky (2008), compensatory consumption can also occur when the individual chooses to consume products that they believe can "mask" some aspect of themselves that they consider to be weak, as a way of improving both their self-image and that transmitted to others. In other words, individuals seek status, flaunting high-value objects that are notoriously associated with a public with high purchasing power. The most easily accessible products are clothing - sometimes replicas - such as glasses, trainers and clothes from certain brands. This behaviour is more commonly found among individuals of lower social and economic status, many of whom have recently ascended to new levels of consumption in Brazil (ABDALLA, 2014, p. 27).

Even in the past, at Rio funk dances in the 1980s - considered the lost decade by economists - some brands were status symbols. In the current decade, in a period of recession, economic conditions and the number of consumers are much higher, as well as per capita income being much higher. As a result, the consumption of status products is much wider and more widespread in society. In the documentary Funk Rio (1994), this point is already made. The large number of people in this context favoured the emergence of rolêzinhos, which are gatherings of poor and peripheral young people in shopping centres.According to Pinheiro-Machado and Scolco (2014), only the combination of variables at different levels - local, international and contemporary - could explain the rolezinhos. The phenomenon of the consumption of luxury brands by socially disadvantaged groups, combined with the ritual of exhibition, is present in the global peripheries, indicating that it is not a phenomenon exclusive to Brazil. The roots of the rolezinhos lie in the contradictions and ironies of modern capitalism, which constantly reinforces in the popular imagination notions of power, distinction and success based on attachment to symbols; generalising this ideal even in groups that are not the target audience (PINHEIRO- MACHADO AND SCOLCO, 2014).

Recently in Brazil, social inclusion policies and rising incomes have somewhat legitimised the model of development based on consumption, a model that previously only covered people from the upper classes; nowadays the opportunities for owning a certain good are much greater. In this way, young people, within a lifestyle characteristic of their generation, end up developing a more ritualised, visceral and exacerbated relationship with the consumerist economic logic. Funk Ostentação is closely linked to the rolezinhos. This style of funk is generally about cars, money and designer labels at exacerbated levels on purpose, externalising the denial of a previously socially defined role. Funk Ostentação is the political discourse that takes place through its positive and creative aesthetic, forging a huge contradiction: the most disadvantaged groups appropriating greater symbols of status and wealth (PINHEIRO-MACHADO AND SCOLCO, 2014).

At the same time, young people affirm their *self* through ostentatious clothing and end up arousing prejudice and apprehension from the upper classes, who understand that their exclusive privilege to certain goods and spaces is under threat. In this logic, prejudice seeks to bring back a state of order in which the "poor are kept in their place" (PINHEIRO-MACHADO E SCOLCO, 2014). It's as if the poor and black people who decades before flaunted their clothes and adornments at funk dances - in theory their territory - left them and invaded shopping centres, which are not "poor people's places". Pinheiro-Machado and Scolco (2014) state that even the high consumption of the emerging classes does not mean that they are seen on an equal footing, suffering from a social reality that is still precarious, in which consumption alone does not resolve the deep historical tensions of social segregation in Brazil.

**The First Contacts**

In August 2015, in order to carry out the research that would make up this monograph, after presenting the research project to Professor Dr Nécio Turra Neto, I found myself with several doubts about the possibility of studying a phenomenon that is part of our city's cultural *underground*, due to its incipience and the limitations it faces. Contact with the people who make up the scene - from the regulars to the MCs and DJs - was essential so that I could analyse the phenomenon as a whole, from the perspective of the public and the artists.

The funk scene in Presidente Prudente has a limited dimension if put into perspective with that of places more associated with the rhythm, such as Rio de Janeiro, São Paulo and the Baixada Santista. Even so, its existence is undeniable, regardless of the difficulties and limitations that make it up. For a few months, I was in contact with the main players in the Prudentina funk scene in order to better understand this phenomenon and its dynamics in the limited context mentioned above.

In November, I administered a questionnaire to 50 young people aged between 14 and 18, students at the Antônio Fioravante de Menezes School and residents of Vila Marina in Presidente Prudente. The answers - which I'll explain below - confirmed my initial suspicion that funk was widespread among young people, as well as the culture of ostentation and the prevalence of the technical scientific information environment.

These questionnaires were tabulated and generated the graphs I present below, in which it is possible to identify/observe that a large proportion of the interviewees enjoy funk and value the relationship between consumption and ostentation.

## 1- You like Funk

**Do you like funk?**

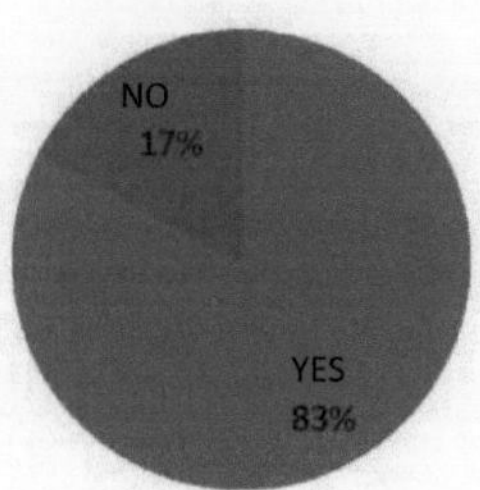


## 2- What's your favourite style of music?

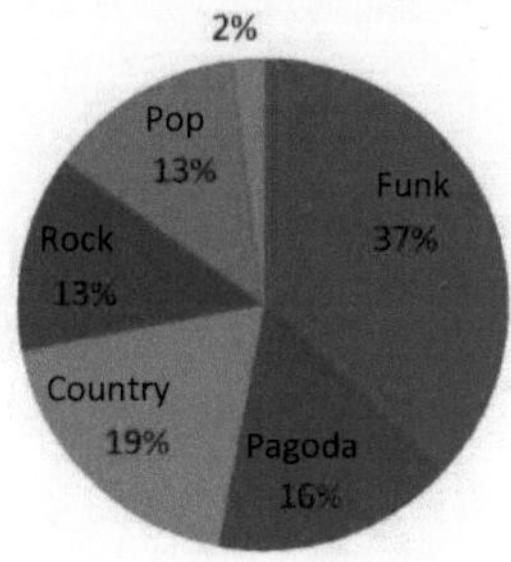


**What's your favourite style of music?** Other

### 3- Do you think music influences listeners?

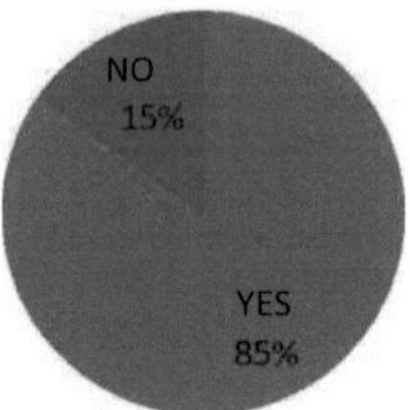



### 4- Do you consider clothes and accessories not very important or very important?

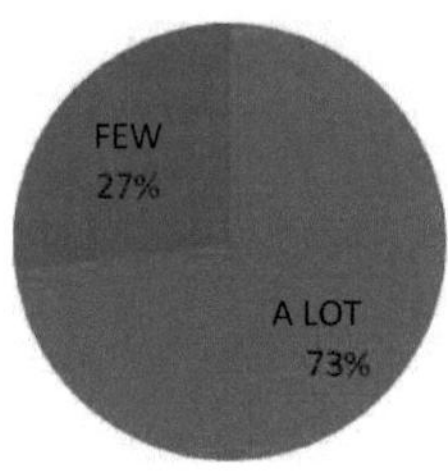



**accessories to
be unimportant or very**

**5- Do you access the Internet on a daily basis?**

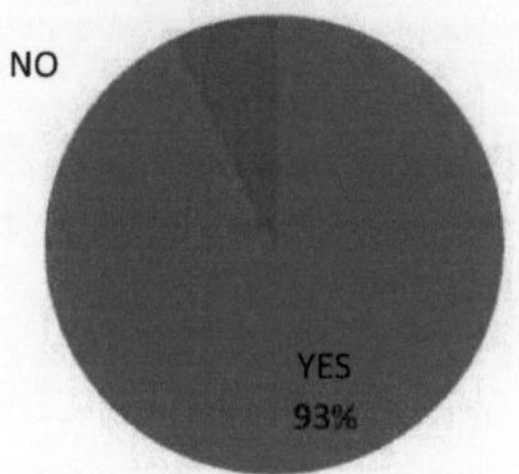


**6- Are you a fan of any artists you met on the Internet?**

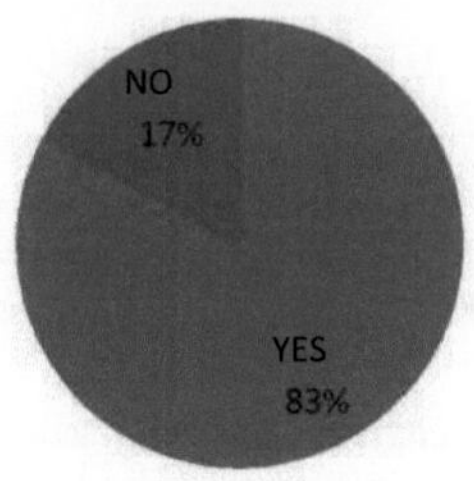

**7- Do you appreciate or admire at least three of the brands listed?**

Hollister; Nike; Abercombie; Lacoste; Adidas; Oakley; Quiksilver

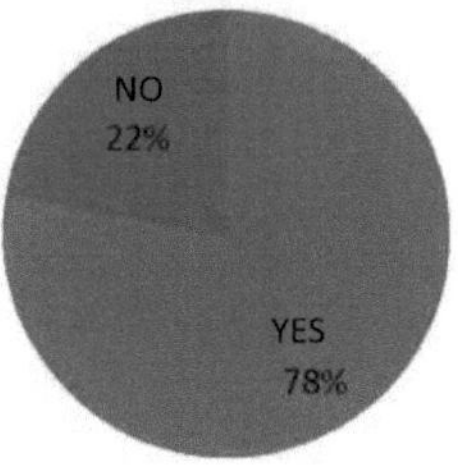


**8- Wearing clothes from these brands -** Hollister; Nike; Abercombie; Lacoste; Adidas; Oakley;

Quiksilver - **passes on an image:**

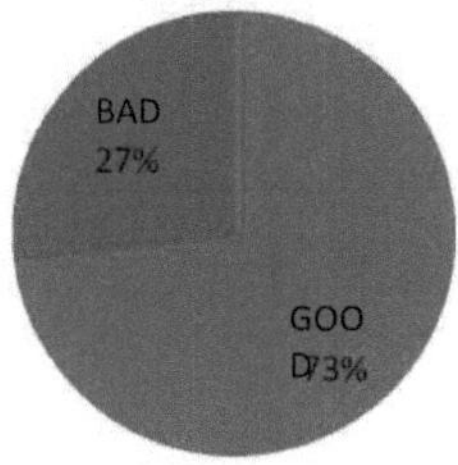


From the perspective of territorial multi-scalarities, the Prudentine scene is a limited scene compared to the national scene, dominated by MCs from Rio de Janeiro, São Paulo and the Baixada Santista. In this way, it is in a geographically distinct territory; even so, it is funk territory due to the networks that are formed based on the technical-scientific environment and are not limited to certain spatial contexts.

The internet and social networks are the main channels for publicising funk and it was through Facebook that I met DJ Robinho and his Tropa. In contact with some of the MCs who make up the troop, I explained my intention to interview them for a *"college assignment"*, the aim of which was to study the funk scene in Presidente Prudente, trying to better understand how this phenomenon, which has little professional support and faces a lot of prejudice in the social sphere, comes about. To my amazement, MC Dark, the first person I met with

When I got in touch, he was very approachable and willing to collaborate. When I saw the MCs' Facebook profiles, I didn't realise that they were all close and part of an artistic collective called Tropa do DJ Robinho. The name is a reference to DJ Robinho, the group's producer, DJ and manager. The group was created out of a desire to strengthen the MCs' careers through collective action. In this way, a show featuring the entire Tropa, or some of its members, would become more marketable and would appeal more to funk lovers, so that they would attend parties, or even as MC Etoo SP said, "Here everyone runs together, if one falls, everyone falls and if one climbs, we brothers climb together".

Once the first contact had been made, I arranged to meet each of the MCs, depending on their availability, so that I could interview them. However, due to the clash between their schedules and mine, we ended up agreeing that I would meet them all together for a group interview, on a date to be decided. In the meantime, I kept in touch by phone with DJ Robinho, probably the most important figure in the Prudentine funk scene, whose profile I'll give below.

**DJ Robinho**

DJ Robinho (Figure 11) is 33 years old, married and has three children. Music came into his life at an early age, in the early 1990s. When he was a child, Robinho already identified with the figure of the DJ and played *scratches* with old vinyls on an old record player he had at home. As a teenager, he started playing with a simple CDJ, but had a series of low-profile jobs as the main way of supporting himself, until the opportunity arose to move to the city of Marília and work for an outsourced concert production team that accompanied artists on tour around Brazil. In this job, Robinho did a bit of everything, from operating sound and lighting boards to setting up the stage. In this role, which is commonly referred to as *roadie*, he acquired considerable *know-how* about music in general and the production of a *concert*. Robinho emphasises that during the six years he has held the position, he has grown not only musically and professionally, but also personally and culturally. What's more, the opportunity to tour with nationally renowned artists such as Charlie Brown Jr, IRA and Kid Abelha gave him the chance to get to know various states in the country, including a stint in Bahia with the famous axé band Terra Samba.

After a few years, for private reasons, DJ Robinho returned to Presidente Prudente with the intention of making a living from his work as a DJ. However, he took on other jobs, such as travelling fruit vendor, as a way of supplementing his family budget. Robinho, who has an impressive knowledge of music, set up a home studio from where he broadcasts the online radio station that bears his name. He also performs at all kinds of events, playing all kinds of music, from weddings, shop openings, church services and funk dances. Noticing the emergence of talented MCs, but lacking in structure and support, DJ Robinho decided to "adopt" these young MCs, providing them with studio time for recording, as well as playing live with the MCs, who are part of his Tropa, and also acting as their manager, being responsible for booking *gigs*.

Having learnt a little about DJ Robinho's life and his relationship with funk, we arranged to meet in a few weeks' time. During this time, I went to a baile funk to make contact with the regulars and do some interviews. It was the first time I'd been to a baile funk in my life. Below is my account of that day, taken from my field diary.

Figure 11: DJ Robinho

DJ Robinho lives with his wife and three children in Jardim Barcelona in Presidente Prudente. At the same address, in one of the bedrooms is the studio (Figure 17) where Robinho records and from where he broadcasts his online radio. The studio is quite simple, in a space of perhaps 10 square metres, with foam acoustic insulation on the walls. The structure surprises me positively, but Robinho says it's *"nothing like a professional studio"*, but agrees that his structure *"does the job"* for the production of Tropa's MCs. I asked Robinho how much a studio similar to his would cost. Robinho couldn't say exactly, but he estimates that it would cost at least five thousand reais - I believe it's more - if you take into account the basic equipment he has.

Robinho uses a notebook accompanied by a CDJ (Figure 13), an MPC (Figure 13) and an MG-32 turntable. A CDJ - basic equipment for any DJ - costs around a thousand reais, although you can find some more basic versions for around half that and others with more features that cost up to six times as much. The MPC is a piece of equipment that, unlike the CDJ, is not every DJ's staple. The MPC is a piece of equipment closely associated with funk, it's an electronic drum machine that allows you to make beats live and it's preferable for a DJ who plays funk to have an MPC piece of equipment that will bring more dynamism to his live performances; an average MPC piece of equipment costs around three thousand reais. As for the MG-32 sound mixer, which controls the sound, it costs around three thousand reais, although there are specific models that can cost ten times as much. The laptop can be any, as long as it has an appropriate sound card, and the voice production and treatment software - known as Pro Tools - can be either free, with free access via the internet, or paid for, with versions that have more resources.

Although I've already mentioned that Prudente's funk scene is part of the lower circuit of the urban economy, it's notable that DJ Robinho is considered "the guy" when it comes to funk in Presidente Prudente. Even with equipment that he himself considers basic, DJ Robinho is still one of the few DJs in the city with MPC equipment. The difficulty of finding a DJ who plays funk in Presidente Prudente is a fact, but even those who do don't have an MPC, which in a way limits their sound, since an MDC greatly increases the sound possibilities as well as live dynamics.

Figure 12: MC Laurinho LB, DJ Robinho and MC Nego at DJ Robinho's home

Figure 13: DJ Robinho's equipment, CDJ and MPC


**Their Night**

On Friday 21 October 2015, at around 11pm, I arrived at Clube Luso, located at kilometre 3 of Rodovia Júlio Budisk, in Presidente Prudente, where the Noite Delas ball took place (Figure 14). Despite my interest in the musical phenomenon called funk ostentação, I had never been to an authentic Baile Funk before. The sky was overcast, threatening rain that didn't come, my basic 2005 Volkswagen car seemed a little out of place in the queue of vehicles waiting to enter the venue, mostly motorbikes - some with altered exhausts to make as much noise as possible - and cars from the 1990s, with alloy wheels with extravagant rims - the so-called orbital wheels. I soon parked my car and headed for the queue to get in. A young man collects my invitation and I enter the hall, without going through any kind of search that could prevent armed people from entering.

I have to confess that, at first, I felt a mixture of concern for safety and satisfaction at finding myself within my field of study - literally - having the opportunity to witness a cultural event that seemed so exotic to me, given my references to fun and partying in Presidente Prudente.

As soon as I queue to get in, I smell a strong aroma of cannabis and discreetly measure the funkers from head to toe. For the boys, the outfits are mainly tac-tel shorts and jeans with logos from the brands: Billabong, Rip Curl and, above all, Oakley, apparently the favourite of the music lovers. T-shirts from Hollister, Abercombie and Quiksilver are also part of the wardrobe, although a considerable proportion opt for polo shirts, mainly from Ralph Lauren, Tommy Hilfiger and, above all, Lacoste. On the feet, Nike and Adidas dictate the tone, while the caps are primarily Quiksilver and the accessories are large, gold watches and gold or silver neck chains, as well as earrings and studs.

The girls, almost without exception, wear high-heeled shoes, short skirts and very tight denim shorts and low-cut blouses or tops with their bellies hanging out. As for accessories and accessories, navel piercings rule, some of them quite extravagant and large. The hair is almost all artificially straightened. For both sexes, orthodontic braces "chavosos[5] " are also a constant.

Everywhere you look you can see smoke, whether from tobacco or cannabis cigarettes, or from hookahs, which are smoked all the time by both boys and girls. It's clear that a large number of the regulars are underage. The music is loud. That night there is no live show, just the DJ playing. Drinks are sold freely. At the bar I buy some catuaba tokens and pour myself the first drink of the night - I've been drinking water all night, after all I still have to drive - I just watch for a while, two girls come up to me and introduce themselves as Jeniffer and Jéssica and say they are cousins; they ask where my friends are, I try to explain that I'm there to do the "field" of an academic research project and that I'm alone, as they don't understand my explanation, I say it's like a newspaper report, to which Jeniffer says: *"But you're not going to say anything bad about the ball, it'll be a drag, then the worms will stick and it'll be over"*. I can assure you that this is not my intention. I ask how old they are. Jessica says she's 18 - although it was obvious she wasn't - Jeniffer hastens to deny it and says they're both 14. Having established this initial contact, I start asking them questions - in the background a chorus plays: *"Novinha vem que tem, novinha, oi vem que tem meu bem, eu to de Camaro e bolsa aquário cheia de nota de cem"* - the girls answer me a few questions and before I can finish asking them they walk off: *"Later we'll exchange ideas uncle, firm, let's go because this sound is crazy".* I had just started playing the music of an MC named Katia, with the simple refrain *"In the art of sex, believe me, I sculpt, I do everything he likes and I still give it to him"*, a sound that I have categorised as Funk Putaria[6] .

The girls disappear into the crowd, I stay there a little longer and go for a walk round the hall. Some time later, Jeniffer and Jessica come walking towards me, from a distance I can already see them pointing at me, along with them a group of a dozen or so people follows, for a moment I'm afraid, what could those girls be pointing at?

As soon as they get close, Jessica says: *"Or fi, you said you were doing an interview bang, right? The boys and girls here want to exchange ideas too, we like funk a lot"*. Before I can answer anything, some of them start talking together: *"Is this a bang for TV Fronteira?" "Is it going to be in the paper?" "I have a cousin who's an MC uncle, a muleque who's really tough, I started to like it because of him". "My brother's in prison..."* Again, as I had done with Jeniffer and Jéssica before, I explain to all of them - five girls and six boys - what I'm doing there and what the aim of my research is, that no one will appear on TV or anything, but that they can look for the finished work in the FCT - UNESP library, where it will be available to the community in general.

Once this contact has been established, I switch on the mobile phone recorder and start asking questions:

- Define Funk for yourself
- How important are these dances and parties?
- Is dressing well and/or wearing certain brands important?
- What do you think about the theme of Funk Ostentação lyrics?
- Does funk suffer from prejudice?
- What do your parents think of funk?

---

[5]Shabby is a synonym for something that attracts people's attention, from a car to jewellery and clothes. In the past it had a negative meaning that has apparently been transformed into an adjective

[6] Funk Putaria or Funk Ousado is a strand of funk characterised by lyrics with an explicitly sexual connotation; sometimes the MC's are young people no more than 12 or 13 years old, which is why they are so popular.

- What did your parents do for a living?
- What is your occupation?

The answers are generally quite similar: some define funk as *"music from the favelas"*, while others say that *"funk is more than that, it's a lifestyle"*. The baile is recognised as a form of expression and that:

> It's very important for those of us who are poor and sometimes if we go to a playboy club like After or Villa Nuts, people look askance at us and for those of us who are underage they won't let us in, but it's not even because of that, it's because we're poor and black (JONATHAN, 16).

As for the clothes, the opinion is unanimous:

> Yes, it's important, fi, how are you going to get to the dance all beaten up and drag a girl? You've got to have the right kit, a cool boot, a Hollister shirt, then the girl will look at you differently, you know? You can already see here who the guys are that attract the most attention by their clothes, outside it's motorbikes and cars, but there are no cars at the dance, so it's the clothes that make you stand out more, you know? (ANDRÉ, 18).

An opinion shared by a girl:

> That's right, you get a flat iron, expensive make-up, put on a pair of shorts and the boys want to come up to you all messed up? That's not good, my friends and I only give morale to boys who are well-groomed, with the right clothes; if they have a car or motorbike, even better (MARIA JÚLIA, 15).

When asked about the themes of Funk Ostentação's lyrics and whether the genre suffers from prejudice, the answers were similar, but with some caveats, mainly regarding prejudice:

> I like the fact that the lyrics talk about getting ahead in life, buying some great clothes, building a house for the family, riding crazy ships; Lon (MC) was just as broke as we are, he has to talk about money, drink and whoring. There is prejudice, we don't listen to funk on 98 or Band radio (GABRIEL, 16).
>
> There is, but I think the prejudice is more because we're black and poor. Give us a lot of money and a car to go to the shopping centre so you can see if we're not treated well by dressing like that, now without money people look at us like we're less (THAINÁ, 16).

I thank them all and tell them that I'll stay there in the same place in the hall for a while longer and that they should ask acquaintances if possible to come to me to answer the same questions, which they do. An hour later, at around 2.30 in the morning when I left, I had done a group interview with 39 people, all of whom had very similar opinions and most of whom were also underage. Although I didn't include an age question, for each interviewee I asked: Are you older or younger? As I was leaving I played MC Rodolfinho's hit - Os mlk é liso-, which begins: *"The story began like this, I saw the mad lives counting money (...) I'm going to visit this shopping centre, buy some Oakley pieces, launch a car called Amarok, a pair of Nike Shox trainers" (MC RODOLFINHO).*

The dance was in full swing, with girls rolling around and dancing nonchalantly, watched by the boys, in corners where kisses and more daring dances were already taking place. During the time I was there, I didn't witness any fights. I left with the impression that a baile funk is no different from a sertanejo concert, for example, as the style of music played is the only one, but the rest is present in both: loud music, drinks, sexually suggestive dances and clothes in character, in the case of a rodeo boots and buckles, at a baile funk caps, shorts and a plethora of Hollisters, Oakleys and Nikes and Adidas everywhere.

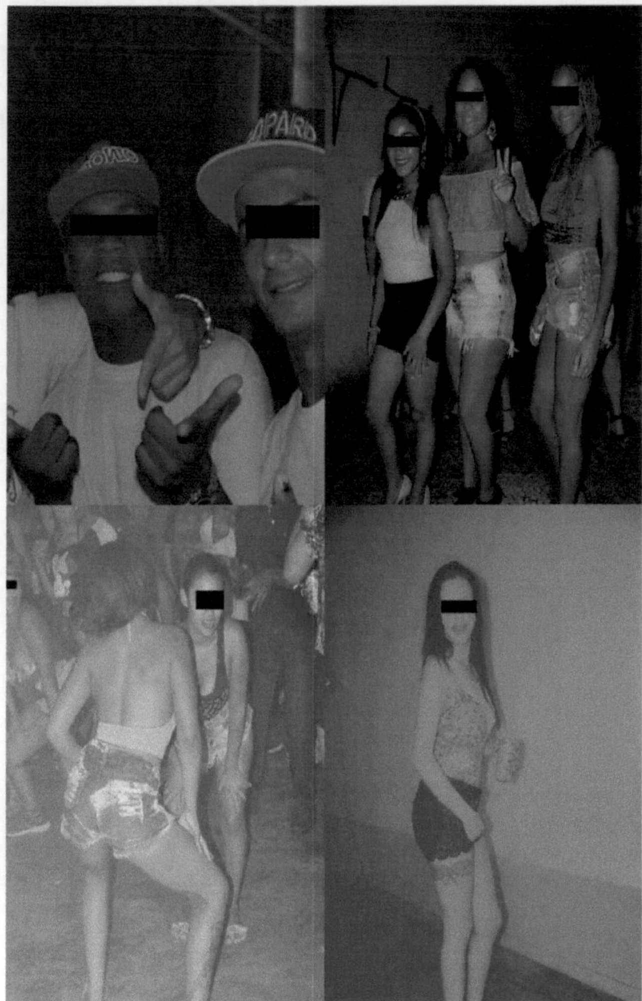

Figure 14: Their Night

Figure 15: Their Night

## Interview with the Troop

A few weeks later, Robinho called me to ask if I was available to meet him and his Tropa that week. We arranged to meet on a Wednesday, but I had to cancel and we rescheduled for Thursday. The place we chose was Alex's Ki- Dogão, a trailer located next to Prudenshopping and frequented by students from FCT-UNESP. Alex, the owner of the trailer that bears his name, is a long-time friend of Robinho's and this helped us to arrange to meet there. Alex's Ki-Dogão is part of a dual context in terms of its territorial characteristics. On the one hand, it is closely related to FCT-UNESP students, who make up a considerable part of its public, and some of the snacks have even been given names that refer to the *Unespian* environment, such as "X-UNESP". It would thus be student territory. However, it is also Tropa's territory, due to the friendship and partnership that exists between Tropa and Alex, which means that the place is frequented by both audiences, in other words, two territorialities inserted in the same context of geographical location.

I'll start by introducing the MCs who make up this Troupe and who took part in the interview:

MC Dark (Figure 14) is 25 years old, has completed secondary school and currently works as a painter during the day. He learnt his trade by doing, as he doesn't have any professional training. MC Dark lives in Álvares Machado with his wife and 5-year-old daughter and says that he became interested in funk when he was 13 and has been involved with music ever since. Dark recognises the difficulties faced by an MC in the local context of Presidente Prudente, given the little space that funk has, yet he says he doesn't think of giving up, out of personal satisfaction and the hope he has of achieving social ascension and, consequently, providing a more comfortable life for his family.

MC Laurinho LB (Figure 14) is 20 years old and hasn't finished secondary school. He did, however, complete a vocational training course as a gas station attendant. Despite his education, Laurinho LB currently works as a leather goods cutter. His interest in funk began around five years ago and for the last three years he has been a duo with MC Nego da Oeste. Mc Laurinho lives with his parents in Presidente Prudente and says that his goal is to be recognised as an MC and provide better living conditions for his family, but he complains about the lack of opportunities to showcase the work that young funk talents find in Presidente Prudente.

MC Etoo SP (Figure 14) is 25 years old, has completed secondary school and has done vocational training as a motorbike mechanic and also as a security guard. However, he currently supports his wife and two-year-old daughter by working as a glazier. Etoo SP had his first contact with funk about six years ago and has been an MC ever since, in the hope of one day being able to help his family financially.

MC Feh is 20 years old and has completed secondary school, as well as a qualification as a general mechanic. He currently works in construction. MC Feh says he has enjoyed funk since childhood and that he became an MC about a year ago. He also says that he is looking for fame and that if one day he manages to earn money as an MC he will *"thank God every day"*.

MC Nego da Oeste (Figure 14) is 22 years old and has completed primary school, has taken a computer course and currently works as a motorbike courier. Nego da Oeste is half of the duo MC Laurinho and had his first contact with funk at the age of 12 and when he turned 18, he followed his dream of being recognised as an MC. MC Nego da Oeste has a daughter and says that his biggest motivation for not giving up his career in the face of difficulties is his family - who he wants to help as soon as possible - especially his daughter.

We had arranged to meet at 9pm, I arrived at around 9.15pm and there were all the MCs and DJ Robinho at a table, sharing a Coca-Cola. I introduced myself and said hello to everyone, the reception was friendly and all the MCs were very approachable, willing to answer anything I asked, always making it clear that they intended to contribute to this research, as well as showing satisfaction at having been the target of an academic study, which aims to document the phenomenon in which they are the main subjects.

Some MCs, such as Nego da Oeste and Etoo, as well as DJ Robinho, took the reins of the conversation, showing a lot of resourcefulness when talking about their work. Little by little, everyone was integrated into the conversation, which unfolded in the most courteous and collaborative way possible. Some points that I consider key to understanding the phenomenon yielded statements that seemed quite pertinent to me, such as when I asked DJ Robinho if he believed that funk suffers from prejudice, to which he replied:

> It does suffer, just like samba in the old days, bro, everything that starts among the poor suffers prejudice, the fuck is that the upper classes end up enjoying it, but they don't admit it, you know? But it's also the funkeiro's fault, the kids go around smoking weed in front of the house, listening to funk, in front of the family, children; then people see it, they associate it, bro, they think funkeiro is all that, then there's the slutty funk too, which delays the funk side even more for families. I myself say, I have my children, I don't want them to listen to funk that talks about whoring, that's wrong. But then you associate it, nobody sees that there are different lines of funk: melody funk, which talks about love; ostentatious funk, which talks about partying, but also about moving up in life, which gives hope to many minors to go after a dream, you know?

to go after a dream, you know, to improve their lives, to help their

mums,

has a brand new car.

Figure 16: Clockwise from top, MC's Dark, Laurinho LB, Etoo and Nego

DJ Robinho's statement is in line with what Sérgio Goldenberg, the filmmaker responsible for the film Funk Rio (1994), said. In a recent interview with the VICE channel, when asked about the prejudice that funk suffers, he replied: "I think the prejudice continues [...] I think there are people who actually go to enjoy the dance, but it's still in the periphery, in the favela. The songs are travelling better, but the event is still a bit segregated."

The Funk Putaria that DJ Robinho refers to is the genre of funk that sings lyrics with explicitly sexual connotations and is adopted by some very young MCs, such as MC Pedrinho, whose hit song is "Dom dom dom" and MC Pikachu, who is only 12 years old, which causes even more controversy.

Some of the MCs are already married with children and believe that some funk lyrics end up *"burning"* funk as a whole in the media, which they already consider to be biased:

> I have a five-year-old daughter and she admires me a lot, she sings all my songs, she knows them by heart, so I even sing funk putaria, but not around her. I know that children end up listening to it, which is why I like ostentatious funk more, because the message talks about partying and also exposes the things we want to have, to hope for, you know? But the media burns it all down anyway, just look at all the bad news that comes out about funk, you know? It's a big struggle for us, you have to love music to keep going (MC DARK).

When I ask him about the limitations of the scene they're in, MC Etoo says:

> The scene here is limited, there's nowhere for us to sing. There'll never be a gig for us in that nightclub in front of the cemetery, you know, on Avenida da Saudade. The people who hang out there even enjoy funk at their parties, but they don't want us playing there. For us, it's just those dances that happen from time to time. And another thing is that we don't have the size of the guys in Sampa or the Baixada, there funk is much stronger, there are lots of places to play shows, you know what I mean? Today, with the internet, the kids can hear these sounds, otherwise there wouldn't even be funk here in Prudente, the internet has revolutionised the business, you know what I mean?

Etoo's statement is similar to that made by Sérgio Goldenberg to the VICE channel, when the latter stated that:

> There's a certain middle-class fascination with the exotic. I have the impression that this happens, but I don't know if the prejudice is over, there is still an exotic view of popular cultural manifestations in Brazil.

Figure 17: DJ Robinho's troop in the studio.

In the same interview, Goldenberg says that consumption and ostentation were already present at funk dances decades before the funk ostentation phenomenon emerged. According to Goldenberg, branded clothes were already considered a status symbol, but on a smaller scale than the one related to ostentatious funk, given that the whole economic context was different, and if today the public at funk dances is made up of a considerable proportion of middle class people, in the 1990s the public was mostly made up of lower class people, who therefore had less purchasing power to consume status items."People put on those clothes with a certain improvisation, something spontaneous, that carioca thing. Shorts, flip-flops, social shirts, football team shirts, branded shorts, branded trainers," said Goldenberg.

Goldemberg's opinion is similar to that of the people I questioned at the Noite Delas ball, almost all of whom confirmed that they believe that clothing brands are important in building an image as an individual, and are, in a way, essential for success with the opposite sex. With regard to consumption and ostentation, MC Feh says that ostentatious funk attracts attention because people look up to MCs who flaunt expensive watches, trainers and clothes. At the time of our interview, MC Etoo, for example, was wearing a gold Invicta watch on his arm; he said it was a replica, worth around RS 120.00. A factory original costs at least R$ 1000.00.

> Style is the main thing for a funkeiro because everyone looks at you, is impressed and also looks up to you, I sing funk ostentação and putaria. And I think brands are important, I like Invicta, Ed Hardy, Mizuno, Nike and team shirts (MC Feh, 2015).

Among MCs, there is a consensus that image and clothes demonstrate success and respect, and it is no coincidence that MCs Daleste, Rodolfinho and Lon are their favourites. This aspect is an example of aspirational consumption and compensatory consumption in terms of trying to convey an image that reflects success and prosperity, according to Rucker and Galinsky (2008). As for the brands they like the most, the MCs cited brands that are already mentioned in various ostentatious funk hits and have become status symbols among funkeiros (NUNES, 2013).

MC Dark says that Oakley, Lacoste, Nike and Adidas are his favourite brands. He explains that these brands are closely linked to ostentatious funk because of their high value:

> Ostentatious funk attracts attention and inspires people, I myself follow my dream of driving a fancy car, having designer clothes and inspiring people to follow their dreams too; besides, people often make you out to be deluded, you know? Funk brings hope.

MC Robinho joins the conversation again, saying that "the question of ostentation" was already present in funk even in the 1990s - corroborating what Guthemberg said - and that funk's essence of "partying and behaving (dressing) well" was present then and today, but today with greater prevalence.

> Funk itself hasn't changed, what has changed are the "new kids" who have come to value clothes much more than before, but that also goes hand in hand with the rise in income, right? Before, the parents didn't have the money to give the kids a pair of trainers, but now they do, you know? (DJ ROBINHO).

Our conversation turns to the Tropa's production and organisational dynamics. DJ Robinho's relationship with the MCs goes beyond the professional and there is a consensus among them that the relationship develops in an even paternalistic way.

> Robinho is a great father to us, he takes care of everything from our production to booking the gigs. The fact that we're not making money yet doesn't mean anything. We're doing what we love and Robinho's support is essential, without him I don't know how it would be, it would be even more difficult (MC ETOO).

MC Etoo brought up a point that I was still going to address, which steered the conversation from that moment on; while they agreed that the figure of Robinho is essential for the existence and maintenance of the scene, they also revealed that the shows are free, they don't get paid. Far from the reality of MCs like Lon, Guimê and Rodolfinho - who earn more than half a million reais a month - the Tropa MCs depend exclusively on the jobs they have during the day, both for their own existence and that of their families.

Funk is considered a passion by the MCs of Tropa, and the dream of being recognised as MCs is something distant and they are fully aware of it. Even so, they consider it a privilege to be able to record their songs in DJ Robinho's studio and then make them available on the Internet, as well as being able to perform live. Due to the intrinsic costs of hiring an MC and an accompanying DJ, many parties end up with just a DJ, with no MCs singing live. MC Robinho is already respected on the city's funk circuit. As a result, he has no shortage of places to play and his fixed fee, according to him, is 200 reais. Adding the MC's from Tropa to his performances was the way Robinho found to move the scene and encourage the development of talent as MC's. It's a kind of barter, the MC's receive all the logistical support from MC Robinho and sing for no pay, which makes the informal and, in a way, amateur relationship between the MC's and DJ Robinho even more explicit:

> There are no MCs who play funk here in Prudente, we don't have a place to play, if you have to hire a DJ then you have to pay to sing, not that we don't practically play. Our scheme is as follows: Robinho strengthens our side, frees up his studio, puts our songs on online radio, books the shows and takes us. Then he earns very little, he earns 200 reais per show, if he pays us it's bad for him, so we sing for free, really, then after the show we go out, have a snack, Robinho pays for us, sometimes a beer, you know what I mean? We all win like this, it's a real family (MC Dark)

Contractual monetary relations in this context end up taking a back seat, since their inclusion would make the funk scene unviable. Thus, the MCs, who are so vocal about ostentation in Presidente Prudente, sing without pay and, although they are part of the funk scene, they are also on the margins of ostentation, and what they sing is much more symbolic than a faithful narrative of what they actually experience, given that when they perform for free, none of the Tropa MCs can give up their paid jobs and join the funk scene in its entirety.

Based on the theory of upper and lower urban economic circuits, Sposito states that there are two urban economic circuits, one would be the upper circuit and the other the lower circuit, both circuits would be defined as: "the set of activities carried out in a certain context, the sector of the population that is essentially linked to it by activity and consumption and would also be defined by differences in technology and organisation" (SPOSITO, 2004, p.187).

Taking this into account, the Presidente Prudente funk scene would be part of a lower economic circuit, since its organisation and technological access don't go much further than what is necessary to maintain the scene itself. The upper local economic circuit does not include the funk scene - see the MCs' statements that they will "never" have the chance to perform in the nightclubs frequented by the wealthier classes - nor does it include it in the upper economic circuit of funk itself, which has its entire production and organisation structure located in other geographical contexts, which do not include the municipality of Presidente Prudente.

In this way, the funk scene in Presidente Prudente is completely inserted in the lower urban economic circuit; even with access to the technology needed to record their songs and access to the technology to publicise them, the MCs have no space to perform and don't receive any payment for them.

**Blackout Night**

On the night of 11 December 2015, Blackout Night (Figure 18), a funk dance, took place at the eclectic Chácara do Sol, already known in our city for hosting a variety of musical events, from pagode shows and funk dances to PVTs (electronic music parties). The frequency of these musical events is high due to its location, close to the asphalt, just behind the SESI-Furquim educational centre, close to Toledo College. The owner of Chácara do Sol, Mrs Vânia - who lives on the edge of the property, in a detached house - says that turnover is high and that hardly a weekend goes by when the Chácara isn't booked for some musical event, with the agenda constantly full.

The event was only publicised on social media and when I told some acquaintances that I was going to a baile funk that night, I heard something like: *"Baile funk here in Prudente? I didn't even know that existed.* The scheduled start time was 23:00 on that hot Friday, the attractions were: DJ Alexx, DJ Juliano Mapão and the main attraction, MC Robinho e a Tropa. The lighting and sound system were provided by Cia. Som from Presidente Prudente and invitations could not be bought in advance, but were only sold at the entrance: women *"free"* all night, men five reais until 01:00 and ten reais afterwards.

In the afternoon, I got in touch with DJ Robinho, as we had agreed, and asked him what time the party would actually be "pumping", since I was going alone and would find the whole Tropa there. Robinho said that he couldn't say for sure, as the dance is very dynamic, but that he would be there from 23:00 and that the crowd should start arriving in larger numbers from 00:00. When I got in touch with MC Dark and MC Etoo, they both told me that unfortunately they wouldn't be able to go to the dance that night because Etoo's 1976 red Passat was having mechanical problems, so I offered to take the MCs to the show, seeing it as an opportunity to continue our conversation and find out a bit more about their lives. Both Etoo and Dark live in Álvares Machado, about 10 kilometres from Presidente Prudente. I arranged to meet Dark after 10pm at the Álvares Machado bus station, from where we would go to pick up Etoo, who lives on a farm by the Júlio Budisk motorway, close to Presidente Prudente, but still within the boundaries of the Álvares Machado municipality.

At the agreed time, I arrived at the bus station and Dark was already waiting for me. We set off to pick up Etoo and, on the way, as soon as Etoo gets into the car, I realise a mistake that he soon notices: "Hey, little brother, where's the funk for us to get into the mood? When we go to a dance, we like to listen to some heavy funk, you know? Like Daleste, mate.

At that moment, my car stereo was playing 98 FM from Presidente Prudente, which was playing the radio hits of the moment. Having forgotten my pen-drive of songs at home, I switch off the car stereo and we start listening to funk on Etoo's mobile phone, which was incredibly loud, first MC Livinho, then MC Lon and so it goes for about 40 minutes, all the way to Chácara do Sol, where we arrive just before 23:00. When we arrive, a few people are already waiting outside, I park my car and we go to the entrance, where there is a male and female security guard, we introduce ourselves as members of

DJ Robinho's Troop and we are let in. The place is still empty and I manage to chat to both the bouncers - who claim to be paid 60 reais for the night's work - and the bar staff - there were three of them - who are paid 50 reais each for their work.

Soon DJ Robinho arrives, accompanied by his wife. We greeted each other and he started setting up his equipment alongside DJ Alexx, the first to play that night. Gradually, people began to arrive and the lights were switched off. The lighting equipment used by Cia. Som surprised me positively, as did the sound, which was very well adjusted, loud but without the hissing characteristic of speakers that aren't in good condition. More and more people are arriving, many underage, and the parade of brands is there, from the queue to the bar or the dance floor. Some wear Oakley Juliet - or Juju - glasses, even though it's late at night. When questioned by me, the 19-year-old who identified himself as Lukas - with a K as he made a point of emphasising - says that:

> It's a bit hard to see, but it's not a problem, bro, it's part of the style, the real Juliet is more than 800 reais, this one is a replica but it costs more than 200, you know? It's expensive too, then the girl sees you wearing it, she sticks with you, you know? (LUKAS, 19)

Lukas said that he works as a clerk at a drinks distributor and that *"every month I earn around 1,100 escudos"*, and that he also works as his uncle's bricklayer's helper on Saturdays. For Lukas, image is very important. He says that he spends almost all his salary on clothes and that *"every Saturday, when I'm not working with my uncle, I go out to buy clothes"*. Lukas was wearing Nike Air trainers - just like the ones I was wearing that day - Billabong shorts and an Abercrombie T-shirt, a Quiksilver cap, Oakley glasses and big sparkly earrings. For a few moments, while I was talking to Lukas, I lost track of the MCs. MC Etoo followed me and called me to join him, MC Nego da Oeste and MC Laurinho LB had just arrived with whisky. There was a bar selling Skol beer - three cans for 10 - vodka, whisky, energy drinks and so on. As no drinks, let alone glass bottles, were allowed in, an exception was made for Nego and Laurinho, who were allowed in with a Red Label whisky and 2 litres of energy drink. I went to the bar, bought several glasses of ice and we prepared our drinks. As I was driving, I intended not to drink, but after some insistence from Nego I ended up accepting a glass. We stood there in front of the DJ who was playing at the moment, DJ Juliano Mapão, who was playing the biggest hits of ostentatious funk: MC Daleste, MC Guimê, MC Rodolfinho.

The dances are in full swing and happen spontaneously, each in their own way, a far cry from the "trams" and rhythmic dancing from one side of the dance floor to the other, as in the 1990s. Most of the girls wear very short denim shorts, the so-called *"short-shorts"*. Some of the more daring girls opt for short skirts or short dresses. I soon realise the trick some girls use to avoid showing their knickers while dancing: a pair of lycra shorts worn under the dress or short skirt in question. At this point, the clock strikes 1.30 and I estimate that around 300 people are already packed into the not very large venue. People continued to arrive and there was already a lot of smoke in the room, although drinks weren't allowed in, hookahs were allowed in and in the corners some people smoked in circles, others smoked ordinary cigarettes and the smell of cannabis permeated the air.

DJ Robinho was getting ready to go live. For a moment, I thought the MCs were going to start singing straight away, but Laurinho LB explained: *"Now Robinho starts playing just to warm up, to get the crowd in the mood, he doesn't even use the MPC much now, then he signals for us to come along, he sends the base on the spot and we sing."*

At that time, around 2am, people stop arriving at the venue. DJ Robinho explains to me that although there is no time for the MCs to start singing, the public who attend the dances already understand that this doesn't happen before 2am and so they choose to arrive a little later, doing the "warm-up" elsewhere, or even going to another party and stopping by the dance at the end of the night, which doesn't cost much, taking into account that you don't pay more than 10 reais to get in.

The entrance fees are similar to those charged for funk dances in the capital and in the Santos lowlands - which demonstrates the targeting of middle- and low-income audiences - even though the Noite do Blackout dance has an inferior structure, whether in terms of lighting or the number of attractions, since some dances, such as those organised and promoted by the São Paulo producer GR6, have up to 50 MCs performing on the same night. On the outskirts of São Paulo, entry to a baile funk costs around 15 reais, based on the fact that it's in the lower economic circuit, but with the insertion of capital, funk is integrated into the upper economic circuit. Entry to a funk show in the affluent Vila Olímpia region of São Paulo, where the majority of the audience is middle or upper class, costs up to ten times more than in the periphery, with nights costing one hundred and fifty reais (MALDONADO, 2015).

According to Juninho Love, executive producer of GR6 - one of the biggest funk producers of the moment - funk today, if managed with company organisation and in a market with demand, can generate a lot of profit and really change lives:

> We have our youth team, the senior team and the stars of the day. In the case of GR6, Pedrinho is Neymar and Livinho is Messi. Today Livinho earns up to R$25,000 per show. And he can do about 30 a month. At 19, he already has two flats and two cars. It's a good start (MALDONADO, 2015, pagination irregular).

Love says that before getting involved in funk he was a backing vocalist in the band of pagodeiro Belo. He saw funk as a business opportunity and has no regrets. Love also says that GR6 produces at least 200 dances a year. It has around 15 MCs on its *roster* (MALDONADO, 2015). In this context, the economic factor is completely integrated, unlike the Prudentine scene and the Tropa MCs, who sing without receiving any remuneration.

For Pereira (2014, pagination irregular):
> Funk ostentação has managed to articulate itself in a very productive and creative way with new technologies. The very well done production of video clips by professionals with a certain specialisation, such as Kondzilla and Funk TV, the management of certain businessmen who sell a set of shows with several MCs to guarantee a whole night's performances in nightclubs, as well as the very entry into this circuit of these clubs have, in a way, guaranteed social and financial returns and mass visibility for this musical genre.

MC Guimê himself, one of the best paid MC's and considered one of the biggest stars of ostentatious funk, has already had to sing for free and in 2013 he received around ten thousand reais per show. About the start of his career, Guimê says the following:

> For a dreamy kid like me, funk worked well. We sang in the favelas on improvised stage tables. Many places didn't even pay. I even did a gig for R$50. My parents and many people disbelieved me. This made me

Amongst the crowd of young people, I notice someone in particular. Cláudia is 40 years old and lives in Cohab in Presidente Prudente. I approached Cláudia and we soon started chatting, noticing that she seemed very excited and couldn't stop dancing. Cláudia told me that she has liked funk since she was a teenager in the late 1980s, and that she still goes to some dances when she can. According to Cláudia, her interest in funk was dormant for a period of her life and resurfaced for a reason:

My daughter Kelly is also here, I only came because she was coming, we always went out together. We leave my granddaughter with my sister and come to the dance. I've been told that I'm not old enough, sometimes I think about it, but I don't know, if I like it and I'm not hurting anyone and I'm still accompanying my daughter, then it's all right (Cláudia., 40, funkeira) Cláudia tells me that her daughter Kelly's father sometimes criticises the fact that they go to funk dances but that, from time to time, he agrees to look after his granddaughter - Kelly's three-year-old daughter - so that they can go out. On other occasions, she leaves the baby in the care of her sister. According to Cláudia, clothes and ostentation were already present in funk from the time she was a teenager, but today the proportion would be greater due to the economic factor:

Clothes? It was already important, at the time it was different, you know, in Cohab at the time there wasn't even tarmac in some streets, so there was a man who had a stereo. This man would put the stereo in front of the house, with those big speakers, you know? Then all the girls and boys from the neighbouring streets would come. You couldn't even call it a dance, but it was crowded, the street was closed, everything was friendly, there was no drug stuff, it was just fun for us. Every Saturday, my sister and I would go to the centre to buy new clothes so we could go out. We'd get money together because we worked in a woman's house, looking after her children. At that time, money was difficult, son. Today, everyone is better off, the minimum wage is low, but it's not as miserable as it used to be. I work at Wilson's drinks, but I give my daughter everything she needs and everything my granddaughter needs, not just me, everyone today has better money, the poor aren't so poor anymore, you know? (Cláudia, 40, funkeira)

Cláudia also mentions that her favourite MC is MC Marcinho and that she doesn't like the sexual appeal of some of the new songs, but that she admires the ostentatious context which, according to her, reflects the current favourable economic context:

Look, a lot of today's lyrics talk about whoring, girls sucking off boys, that sort of thing. I don't see a problem with playing them at a dance, but you can't play them on your home stereo at high volume, there are other children in the neighbourhood who can hear them and that's not cool, my daughter herself doesn't listen to them around my three-year-old granddaughter at home. What I like is the ones that talk about cars, jewellery, you know? Today the poor can have things here in Brazil, back then

when my sister and I were little, we couldn't even dream of having these branded things. I voted for Lula, I voted for this Dilma too, with her things are getting a bit worse, but I vote for these PT people. When Dilma became president I thought she was Lula's wife, Kelly told me she wasn't, but I voted anyway; see if there was a Bolsa Família before? See if there was a way to go to university? Kelly is there, she wants to go into administration and God willing she will (CLÁUDIA, 40, funkeira).

Some time later, DJ Robinho announces that the MCs are ready to start, the first being the duo formed by MC Nego and MC Laurinho LB. Then it was MC Etoo's turn and, finally, MC Dark closed the MC night, before DJ Taty took over to end the event. The entrance of the MCs caused a stir and everyone crowded right in front of the space reserved for the MCs to sing. Each MC sang three songs, not exceeding twelve minutes each, which led to shouts of "keep going, keep going, keep going". DJ Robinho demonstrated total mastery of the art of disco throughout the MCs' performance, making live beats using MPC equipment. After MC Dark finished his set, DJ Taty took over the disco for half an hour and the dance ended at around five in the morning. The people gradually dispersed, I had a brief word with DJ Robinho and left with MC Dark to take him back to his house in Álvares Machado, at the same time as Dona Vânia - the owner of the farm - was already picking up rubbish from the floor. After all, around midday that Saturday, the new tenants for the weekend would be arriving and, due to cost-cutting, no one had been hired to clean up the place.

Figure 18: Blackout Night

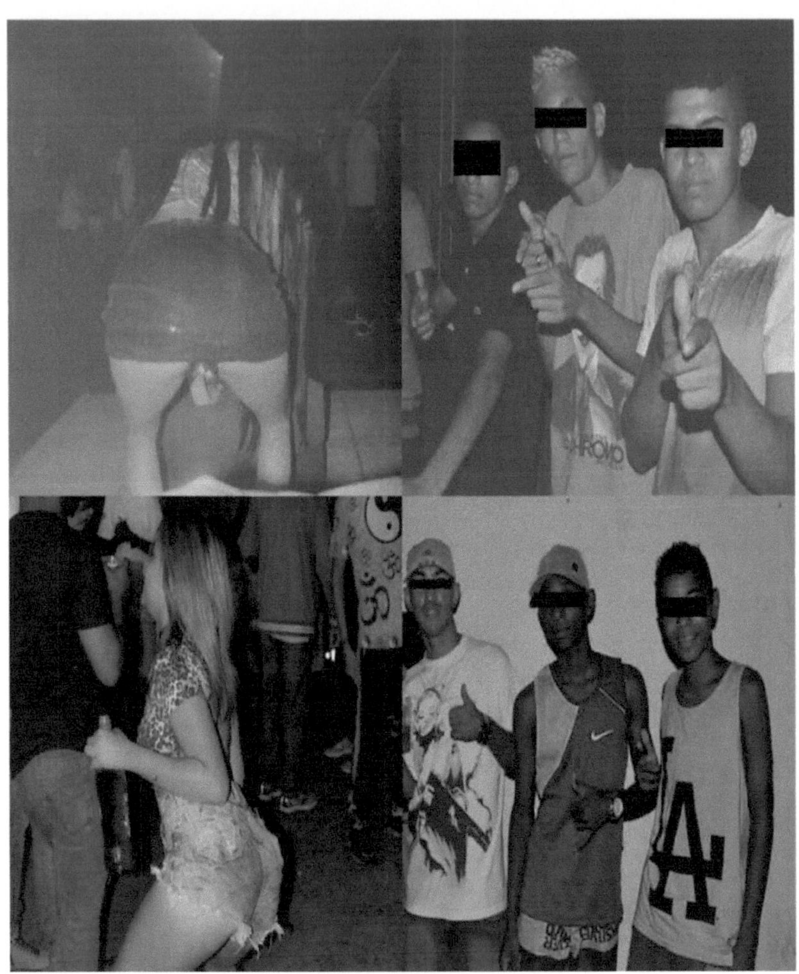

Figure 19: Blackout Night

# CHAPTER 6 FINAL CONSIDERATIONS

The research carried out leads us to believe that the Funk scene in Presidente Prudente is limited when compared to that of São Paulo and Baixada Santista - excluding the Rio de Janeiro scene, which is the most traditional and structured in the country - but its existence is undeniable and its maintenance is closely related to the prevalence of the technical scientific information environment.

Located in the far west of São Paulo, the municipality of Presidente Prudente is in a different geographical and cultural context from the one that gave rise to Funk Ostentação; while in the city of São Paulo and the Baixada Santista there are a considerable number of favelas and irregular occupations, in Presidente Prudente they don't exist. In Presidente Prudente there are deprived areas, but they are far from the universe of a favela. The similarity between the aforementioned localities lies in the fact that they have a considerable number of people who used to belong to the less favoured classes of society and who have ascended to the consumer market in the last decade, becoming de facto consumers. This consumption was initially aimed at basic needs such as housing, for example. However, once these demands had been met, the desire arose among young people from these families who had risen socially to consume high-value products that are status symbols, more associated with the upper classes.

Whether in Presidente Prudente, São Paulo or the Baixada Santista, young people from these families who have risen socially have the same desires for consumption and personal expression, even though they are in different urban contexts. In this way, the songs of the Funk Ostentação genre have their message absorbed by young people from the "new middle class", or Class C young people, in different locations. It can be seen that in the locations where Funk Ostentação has emerged, there is a wide range of related options, either through the prevalence of nightclubs, which promote Funk shows, or through parties held in the streets, the so-called fluxos. In Presidente Prudente, the options are limited, taking into account that the city's nightclubs don't offer space for Funk-related events, nor do the radio stations play music from the genre.

In carrying out this work I was able to be in direct contact with fans and MC's and DJ's of Funk Ostentação. It became clear that the scene's existence is directly linked to their personal desire - mainly for MCs and DJs - since there is no structural support from any production company - whether or not it's directed at Funk - or nightclubs or radio stations. As a result, the Internet and social networks are their only means of promotion and the MCs themselves, driven by the desire to showcase their work and talent, end up singing for no pay. This is in stark contrast to the situation of nationally famous MCs with a whole structure, such as MCs Guimê and Lon, who earn up to six hundred thousand reais a month, performing up to fifteen shows every weekend.

The economic improvement of society in general over the last decade is a phenomenon that covers the whole country, but the uniformity of the Funk Ostentação scene is not a reality. Fans are indeed spread throughout Brazil, but the centres of production and influence are still São Paulo and the Baixada Santista, with all other local scenes being fairly limited.

Among the MCs I came into contact with during the course of this research, there was a clear sense of satisfaction at being able to demonstrate their talent as MCs. However, they were also uncomfortable with not being recognised as artists or being paid, as they had to work other jobs during the day to support their families. Finally, there is the doubt and the suggestion of future research into the long-term duration of the local

Funk scene in Presidente Prudente, given that both the MCs and DJ Robinho - described by themselves as a "paizão" - perform their duties within Funk much more out of appreciation for the music than for financial results, which can become unsustainable due to the time required without pay. I believe that if DJ Robinho and his Tropa leave the scene, funk dances will still take place in the city, but without the presence of MCs singing live; which would weaken the scene, since original songs would be left out, replaced by some DJ playing hits by nationally famous MCs on his CDJ.

The future of the funk scene in Presidente Prudente is uncertain, but the situation suggests that the audience to be explored is broad, considering that among the fifty young people aged between 14 and 18 who answered the questionnaire applied in this survey, 83 per cent said they enjoyed funk and 37 per cent said it was the style of music they liked best. This hypothetical situation depends on the structuring and insertion of capital into the scene, as in São Paulo and the Baixada Santista, where stories told in the lyrics of MC Lon or MC Rodolfinho have become concrete reality and serve as inspiration for other young people who have talent and are thinking of embarking on an artistic career with visibility, in which they can ascend socially in a concrete and not just poetic way.

# REFERENCES

ABDALLA, Carla. C. **Rolezinho pelo funk ostentação:** a portrait of the identity of young people from the periphery of São Paulo, 2014. Dissertation (Master's) - Getúlio Vargas Foundation - FGV, São Paulo, 2014

AMENDOLA, Beatriz. **More liked than celebrities, anonymous people become famous on the web.** 2014. Available at :
<http://celebridades.uol.com.br/noticias/redacao/2014/11/13/mais-curtidos-do-que-celebrities-anonymous-become-famous-on-instagram.htm> Accessed on: 23 July 2015

ARAÚJO, Gisele. **Facebook celebrities**: young miners gain notoriety and money. 2013. Available at: <http://www.hojeemdia.com.br/horizontes/os-famosinhos-do- facebook-jovens-mineiros-ganham-notoriedade-e-dinheiro-1.193462> Accessed on: 05 Jul. 2015

BATISTELLA, Camila. **Consumption and indebtedness in the Brazilian middle class at the beginning of the 21st century.** 2014. Monograph (BA) - University of Brasília - UnB, Brasília, 2014

CHAGAS, Anivaldo. T. R. **O questionário na pesquisa.** 2000. Fundação Escola de Comércio Álvares Penteado - FECAP, São Paulo, 2000

COLDIBELI, Larissa. **Ostentatious funk label seeks Abercrombie customers with shirts for R\$200.** 2014. Available at :
<http://economia.uol.com.br/empreendedorismo/noticias/redacao/2014/05/20/grife-de-funk- ostentacao-busca-clientes-da-abercrombie-com-camisas-a-r-200.html>. Accessed on: 13 June 2015

COLOGNESE, S. A.; MÉLO, J. L. B. de. The interview technique in social research. **Cadernos de Sociologia**, Porto Alegre, v. 9, p. 143-144, 1998

DAYRELL, Juarez. **Music enters the scene**: rap and funk in the socialisation of youth in belo horizonte. 2001. Thesis (Doctorate) - University of São Paulo - USP, São Paulo, 2001.

DEMATTEIS, Giuseppe. Local Territorial System (SLOT): an instrument for representing, reading and transforming the territory. In: ALVES, Adilson. F.; CARRIJO, Beatriz. R.; CANDIOTTO, Luciano. Z. **Territorial Development and Agroecology**. São Paulo: Expressão Popular, 2008. p.34.

DIÓGENES, Glória. **Cartographies of Culture and Violence** - gangs, galleys and the hip hop movement. São Paulo: Annablume, Fortaleza: Secretariat of Culture and Sport, 1998.

EGUENDEL, A. **Sam Cooke's civil rights anthem a change is gonna come**. 2014. Available at: <https://musicsoftheworld.wordpress.com/2014/03/11/sam-cookes-civil-rights-anthem-a-change-is-gonna-come/>. Accessed 12 Nov. 2015

FIGUEIREDO, Marina. D. de; CEVEDON; Neusa. R.; & SILVA; Alfredo. R. L. da. The devaluation of social groups in the common space of small organisations: a study of social representations in a shopping centre. **Organizações & Sociedade**, Salvador, v. 20, n. 64, p.14, jan. 2013.

GIL, Antonio. C. **Methods and techniques of social research.** São Paulo: Atlas, 1987

HAESBAERT, Rogério. Dilemma of concepts: space-territory and territorial containment. In: SAQUET, Marcos. A.; SPOSITO, Eliseu. S. (Org.) **Territories and territorialities**: Theories, processes and conflicts. São Paulo: Expressão Popular. 2009, p. 95-118.

GOMBATA, Marsílea. **Without social criticism, ostentatious funk falls in favour with**

**the middle class**. 2013. Available at: http://www.cartacapital.com.br/cultura/sem-critica-social-funk-__de-ostentacao-cai-no-gosto-da-classe-media-1321.html. Accessed on: 14 Nov. 2015

HAESBAERT, Rogério. Territory and multi-territoriality: a debate. **GEOgraphia**, Rio de Janeiro, year IX, n. 17, p. 19-45, 2007.

______. **O mito da desterritorialização**: do "fim dos territórios" à multiterritorialidade. Rio de Janeiro: Bertrand Brasil, 2004. p.43.

HALL, Stuart. **A identidade cultural na pós-modernidade** (5 ed.) Rio de Janeiro: DP&A, 2001

HERSCHMANN, Micael. **Shaking up the 90s**: funk and hip-hop, globalisation, violence and cultural style. Rio de Janeiro: Rocco, 1997.

HOLLANDA, Sérgio. B. **Roots of Brazil**. Rio de Janeiro: José Olympio, 1936.

KNUTSE, Kristian. **Otis Redding at The Factory: One night only in Madison**. 2007. Available at <http://isthmus.com/arts/otis-redding-at-the-factory-one-ni__ght-only-in-madison/>. Accessed on: 14 December 2015.

KREPP, Ana. **'Rolezinhos' emerged with young people from the periphery and their fans**. 2014. Available at: <http://www1.folha.uol.com.br/cotidiano/2014/01/1397831-rolezinhos-surgiram-com-__jovens-da-periferia-e-seus-fas.shtml>. Accessed on: 08 July 2015.

McLUHAN, Marshall. **The Media as Extensions of Man**. São Paulo: Cultrix, 2005.

MEDEIROS, Janaína. **Funk carioca:** crime or culture? Rio de Janeiro: Terceiro Nome, 2006.

MOREIRA, D. **Consumption on the outskirts of São Paulo grows more than in the centre.** 2011. Available at: <http://exame.abril.com.br/pme/noticias/consumo-naperiferia-de-sao-paulo-__grows-more-than-in-the-centre> Accessed on 15 Jul. 2015

NEAL, Mark A. **We remember                         SamCooke.**       2013. Available at: <htttp://www.thenewblackmagazine.com/view.aspx?index=3040>. Accessed on: 15 Dec. 2015

NERI, Marcelo. C. S. **A nova classe média:** o lado brilhante dos pobres. Rio de Janeiro: FGV/CPS, 2010.

NERI, Marcelo. C. S. **Overcoming poverty and the new middle class in the countryside**. Rio de Janeiro: FGV/CPS, 2012.

NUNES, Gustavo. R. **The explosion of symbolic power and signs**. 2013. Available at: <http://observatoriodaimprensa.com.br/diretorio-__academico/ed749 a explosao do poder simbolico e de signos/>. Accessed: 14 July 2015

                    Available at      : <http://origemdapalavra.com.br/site/palavras/identidade>. Accessed 13 Nov. 2015

PADILHA, Valquíria. **Shopping Centre**: the cathedral of merchandise. São Paulo: Boitempo, 2006

PINHEIRO-MACHADO, Rosana; SCALCO, Lucia. M. Rolezinhos: Brands, consumption and segregation in Brazil. **Revista de Estudos Culturais**, São Paulo, n.1, irregular pagination, 2015. Available at:

PEREIRA, Alexandre. B. Ostentatious Funk in São Paulo: imagination, consumption and new information and communication technologies. **Revista de Estudos Culturais**, São Paulo, n.1. paginaçãoirregular            ,                  2015.
                    Disponívelem       : http://www.each.usp.br/revistaec/?_q=revista/1/funk-ostenta%C3%A7%C3%A3o-em-

s%C3%A3o-paulo-imagina%C3%A7%C3%A3o-consumo-e-novas-tecnologias-da-informa%C3%A7%C3%A3o-e-da, Accessed January 2016.

RECUERO, Raquel. **Social Networks on the Internet, Information Diffusion and Journalism:** elements for discussion. Santa Cruz do Sul: UNISC, 2009

SALLES, Ecio. P. de. The good and the ugly: forbidden funk, sociability and the production of the common. **Z Cultural Magazine**. Federal University of Rio de Janeiro - UFRJ, 2014. Available at: <http://revistazcultural.pacc.ufrj.br/o-bom-e-o-feio-funk-proibidao-sociabilidade-e-a- producao-do-comum-de-ecio-p-de-salles/ Accessed on: 25 Nov. 2015

SANTOS, Milton. **The nature of space**: technique and time, reason and emotion. São Paulo: Editora da Universidade de São Paulo, 1996

SOUZA JÚNIOR, Carlos A. da. C. S. **Expressions of exhibitionism on Facebook**: an analysis based on the Social Sciences, 2014. Dissertation (Specialisation) - University of São Paulo - USP, São Paulo, 2014

SOUZA, Jessé. The underside of Brazilian society. **Revista Interesse Nacional**, Piracicaba, v.14, p. 33-41. 2011/2012

SOUZA, Marcelo J. L. de. Territory: on space and power, autonomy and development. In: CASTRO, I. E. de; GOMES, P. C. da C.; CORRÊA, R. L. (org.).**Geografia**: conceitos e temas. Rio de Janeiro: Bertrand Brasil, 1995.p.77-81

SPOSITO, Eliseu S. **Geografia e filosofia**: contribuição para o ensino do pensamento geográfico. São Paulo: Editora da UNESP, 2004.

STAMBOROSKY, Amauri. **From James Brown to Rap das Armas, see the timeline of carioca punk**. Globo, 2009. Disponível at:<http://g1.globo.com/Noticias/Musica/0,,MUL1288890-7085,00-FROM JAMES BROWN TO THE RAP OF WEAPONS SEE THE LINE OF TIME OF F UNKCARIOCA.html. Accessed on: 12 Oct. 2015

TERRA. **MC Daleste is the celebrity with the highest number of searches in 2013**. 2014. Available at: http://tecnologia.terra.com.br/internet/mc-daleste-e-o-famoso-com-maior-alta-de-buscas-no- google-in-2013,1d3e2dbbd5df2410VgnVCM3000009af154d0RCRD.html Accessed on: 17 Oct. 2015

TURRA NETO, N. **Múltiplas trajectórias juveniles em Guarapuava**: territórios e redes de sociabilidade. 2008. Thesis (Doctorate) - Universidade Estadual Paulista - UNESP, Presidente Prudente, 2008.

VIANA, Luciana. R. **O funk no Brasil**: Música desentermediada na cibercultura. SONORA.
Campinas, v. 3, n. 5, 2010.

VIANNA, Hermano. **O Mundo Funk Carioca**. Rio de Janeiro: J. Zahar, 1987.

WHELIE, Papa. Miami Bass: The Primer. STYLUS Magazine, 2007. Available at: <http://www.stylusmagazine.com/articles/weekly article/miami-bass-the-primer.htm>.
Accessed on: 08 Nov. 2015

**ANNEX:**
**MC Feh**
**Recalc**
Ever since I was a child, I had to keep quiet and
today the whole nation, I stop and I am remembered
today's recalcitrance is me feeling sorry for myself
women and showing off, that's my dilemma.
the Juju is on the face and the Nike gel is on the foot
you even know who you are from the ship.
this is the crazy life, and I pull on the boot
if you don't like it then please turn your back
come and rock us at the ballad
and our fame will reach the capital
we're in the mood to rock and roll
and God willing it will be a national top.
**MC Etoo SP**
**I prefer my faithful**
The new girl's crazy, she's looking me up and down
Give me a little nibble on the mouth, hi there, I'm done.
Relax, you naughty little girl, watch the tape and I'll show you the view
Don't come and play the shoveller because I'm not a maloqueiro.
I'm not into mistresses, my faithful one is fine with me
If you keep pushing me, I'll catch you and you'll regret it
She gets crazy, she even steals my mind
You've got your eye on my pendants!
Then she goes off saying that men are no good
Look at what he's saying about us
I prefer my faithful and I don't care about lovers
The faithful close with me and the mistress wants gold and diamonds
**MC Etoo SP**
**Our tram is fucked**
Our tram is fucked up, recalcitrant people attack us because we're in fashion
Our tram is fucked up, recalcitrant people attack us because we're fashionable
The R1 gives us a wave, the clown gives us a hard time
A Camaro convertible is just the way they like it,
With various kits, launches, fashion
Ed Hardy, Brooksfield, 212 and Polo play
And at nightclubs we're called king
Call all the young girls, our tram likes it
And to close with us, they're even making a bet
Bet your heart out, thinking you'll win
But if I catch you today, I won't even remember tomorrow

Our tram is a monster, DJ, you know that?
Talent like ours is unlimited
He's just as authentic as Daleste said
Not authentic, authentic
When I'm in the spinning top I see the drool of a sucker
Thinking I work in a bank or am an entrepreneur
My pocket is an aquarium, look at the thickness of the cord
It's giving me a kink, it's dragging on the ground
But it's not my gold, it's the drool that's fallen
Take your envy and go to hell
Our tram is fucked
Our tram is fucked up, recalcitrant people attack us because we're in fashion
Our tram is fucked up, recalcitrant people attack us because we're in fashion

**MC Dark**

**King of the Box**

That's enough
King of the box, full of suitcases
Close with the firm, you can stick it, get involved
King of the cabin, my firm is strongMounted on a Veloster, our tram can
We count on God, but we don't count on luck Chrome wheels, thousand-cylinder motorbikes
Bums stare, but don't understand anything Pique we call ourselves, we're in the car Novinha ta maluco, so drag it on the ground
She's had Chandon, she's mad, she's crazy
Come to the cabin, it's the madmen of the neighbourhood
So come along, come along, come and make a mess
And at the end of it all, I know where it will end up
King of the box, full of suitcases
Close with the firm, you can stick it, get involved
King of the cabin, my firm is strong
The motto is those who can, can, those who can't shake it Full of fish in their pockets, pique moleque do torro Diamond studded around pure gold
Armani pulled Ferrari, Ed Hardy's shirt
Okley, Polo play, designer, Lacoste, Aeropostale
King of the box, full of suitcases
Close with the firm, you can stick it, get involved
King of the cabin, my firm is strong

# Table of contents

Buy your books fast and straightforward online - at one of world's fastest growing online book stores! Environmentally sound due to Print-on-Demand technologies.

Buy your books online at
**www.morebooks.shop**

Kaufen Sie Ihre Bücher schnell und unkompliziert online – auf einer der am schnellsten wachsenden Buchhandelsplattformen weltweit! Dank Print-On-Demand umwelt- und ressourcenschonend produzi ert.

Bücher schneller online kaufen
**www.morebooks.shop**

Printed by Books on Demand GmbH, Norderstedt / Germany